Technology Transfer

A
Communication
Perspective

edited by

Frederick Williams
David V. Gibson

SAGE PUBLICATIONS
The International Professional Publishers
Newbury Park   London   New Delhi

Copyright © 1990 by Sage Publications, Inc.

For information address:

SAGE Publications, Inc.
2455 Teller Road
Newbury Park, California 91320

SAGE Publications Ltd.
6 Bonhill Street
London EC2A 4PU
United Kingdom

SAGE Publications Pvt. Ltd.
M-32 Market
Greater Kailash I
New Delhi 110 048 India

Printed in the United States of America

Library of Congress Cataloging-in-Publication Data

Main entry under title:

Technology transfer: a communication perspective / edited by
   Frederick Williams and David V. Gibson.
       p.   cm. —
   Includes bibliographical references.
   ISBN 0-8039-3740-7 (C). — ISBN 0-8039-3741-5 (P)
   1. Technology transfer.   I. William, Frederick, 1933-   .
II. Gibson, David V.
   T174.3.T3757   1990
   338.9′26—dc20

                                                        90-8778
                                                        CIP

FIRST PRINTING, 1990

Sage Production Editor: Astrid Virding

Contents

Acknowledgments   7

Introduction
Frederick Williams and David V. Gibson   9

I. Challenges of Technology Innovation and Transfer
1. The Coming Economy
George Kozmetsky   21

II. The Organizational Setting of Technology Transfer
2. The Intraorganizational Environment:
Point-to-Point Versus Diffusion   43
Dorothy Leonard-Barton

3. The Interorganizational Environment:
Network Theory, Tools, and Applications
Ellen R. Auster   63

III. Contexts of Technology Transfer
4. Research Consortia: The Microelectronics and
Computer Technology Corporation
Christopher M. Avery and Raymond W. Smilor   93

5. University and Industry Linkages:
The Austin, Texas, Study
G. Hutchinson Stewart and David V. Gibson   109

6. University and Microelectronics Industry:
   The Phoenix, Arizona, Study
   Rolf T. Wigand   132

7. New Business Ventures: The Spin-Out Process
   Glenn B. Dietrich and David V. Gibson   153

8. Transfer via Telecommunications:
   Networking Scientists and Industry
   Frederick Williams and Eloise Brackenridge   172

IV. International Perspectives
9. Mexico and the United States:
   The Maquiladora Industries
   Eduardo Barrera and Frederick Williams   195

10. Japan: Tsukuba Science City
    James W. Dearing and Everett M. Rogers   211

11. Italy: Tecnopolis Novus Ortus and the EEC
    Umberto Bozzo and David V. Gibson   226

12. Bangalore: India's Emerging Technopolis
    Arvind Singhal, Everett M. Rogers,
    Harmeet Sawhney, and David V. Gibson   240

13. Multinationals: Preparation for
    International Technology Transfer
    Eun Young Kim   258

V. The Literature of Technology Transfer
14. The State of the Field: A Bibliographic View
    of Technology Transfer
    David V. Gibson, Frederick Williams,
    and Kathy L. Wohlert   277

Index   293

Acknowledgments

Many people and organizations helped make this book possible. First, we are indebted to Dr. George Kozmetsky, Director, and Dr. Raymond W. Smilor, Executive Director, of the IC^2 (Innovation, Creativity, and Capital) Institute at the University of Texas at Austin. Several of the research projects reported in the following chapters were conducted with the assistance and support of the IC^2 Institute.

We also are indebted to the members and many contributors to the Technology Transfer Research Group at the University of Texas at Austin. The multidisciplinary research group was formed in 1987 and is sponsored by the Center for Research on Communication Technology and Society, College of Communication, the Bureau of Business Research, College and Graduate School of Business, and the IC^2 Institute. Former and current graduate students Christopher Avery, Eloise Brackenridge, Richard Cutler, Glenn Dietrich, Eun Young Kim, Harmeet Sawhney, Gary Stewart, and Francis Wu have contributed much to our thinking. A special note of thanks to graduate student and colleague Kathy Wohlert, who was of immeasurable help in literature searches, brainstorming sessions, and general good humor in helping to keep the book on track. This book reads as well as it does, in large part, due to her exceptional proofreading, editing, and indexing. We also express our gratitude to Ralph Segman, editor of the *Journal of Technology Transfer,* James M. Wyckoff, past president of the Technology Transfer Society, and F. Timothy Jahis, current president of the

Technology Transfer Society, for their support of and association with the activities of the Technology Transfer Research Group.

We owe special thanks to Connie Crytzer of the Center for Research on Communication Technology and Rose R. Orendain and Linda Teague of the IC^2 Institute for their assistance with typing this manuscript and preparation of the figures and tables.

We greatly appreciate the support of our deans, Robert Jeffrey of the College of Communication and Robert E. Witt of the College and Graduate School of Business, as well as Professor James S. Dyer, Chairman of the Department of Management Science and Information Systems.

Finally, we wish to acknowledge the authors of each of the following chapters. Their writings collectively make *Technology Transfer: A Communication Perspective* an important contribution to an improved understanding and to the scholarship and practice of technology transfer.

—Frederick Williams
David V. Gibson

Introduction

FREDERICK WILLIAMS
DAVID V. GIBSON

In this opening chapter, the editors present a rationale for studying technology transfer as a communication process. Frederick Williams is Director of the Center for Research on Communication Technology and Society at the University of Texas at Austin, where he also occupies the Mary Gibbs Jones Centennial Chair in Communication, and is the W. W. Heath Fellow in the IC^2 Institute. Williams is author of about 70 articles and 30 books on topics in communications and information technologies. He served earlier as founding dean of the Annenberg School for Communication at the University of Southern California. David V. Gibson is a member of the faculty of the Graduate School of Business and is a Research Fellow in the IC^2 Institute at the University of Texas at Austin. He is author and editor of books and author of articles on technology transfer, economic development, and the management of information systems. Williams and Gibson codirect the Technology Transfer Research Group at the University of Texas at Austin, cosponsored by the Center for

Research on Communication Technology and Society in the College of Communication, the Bureau of Business Research in the Graduate School of Business, and the IC^2 Institute.

Technology Transfer as a Communication Process

Technology transfer is discussed and analyzed increasingly in current business, government, and academic publications (Reich, 1989). It is the focus of economic development conferences and is beginning to appear in academic course catalogs. Although the application of the term takes on specialized meanings in government, law, science, or economic development, there is a common theme in its usage. In a broad view, technology transfer reflects all or some components of the process of moving ideas from the research laboratory to the marketplace. In brief, technology transfer is the application of knowledge.

In our view, borne out of three years' experience with a faculty-student-practitioner research group at the University of Texas and various conferences and research meetings, we have focused on the process of technology transfer as a communication phenomenon. To make technology transfer successful requires overcoming the many barriers to communication encountered when individuals use different vocabularies, have different motives, represent organizations of widely differing cultures, and when the referents of the transactions may vary from highly abstract concepts to concrete products. At the heart of the matter, we consistently observe these problems in such communication transactions between individuals (such as scientist-scientist, scientist-client, manager-customer, manager-scientist, etc.), within and among corporations, in university-industry collaboration and new R&D consortia, between government and industry, and in international transfer. (In communication research literature, this often is referred to as a general *source-destination* paradigm.)

These challenges also exist in the more implicit technology transfer activities embodied in managing transnational or multinational operations, setting up corporate spin-offs, or in the continuing education of scientists. From the vantage point of those of us who are researchers in this area, we see technology transfer as an especially rich and challenging field for communication research. It is an area that may simultaneously call for group, organizational, or interpersonal communication research strategies, as well as an examination of the role the new media are playing in the technology transfer process (as in the many telecommunications and data networks referred to in subsequent chapters).

Discussions of technology transfer are apt to run the gamut from broad charges that U.S. competitiveness depends upon our success in bringing inventions to the marketplace to equally broad accusations that we are not teaching the nation's youngsters enough about the creation and management of technology so they can compete in the global job market. In a strategic vein are debates about research consortia and other strategic alliances where traditional free-enterprise competition among companies is traded for intra-industry collaboration. To this is added the pros and cons of seeking federal support and direction so that the challenges hurled by Japan, Inc. might be countered with a U.S.A., Inc. On a more detailed level are questions about scenarios concerning joint venturing, licensing, timing of legal actions like patents and disclosures, or how to get laboratory and university researchers to communicate to the world of business (and vice versa). Although much is being written as well as discussed in conferences and workshops on enhancing the technology transfer process, comparatively less has been done on a more conceptual level of analysis. The latter is the goal of this volume, in which we advance the particular view that it is valuable to examine technology transfer as a communications process—generally, to view its many aspects using this common perspective.

General Questions

In some respects, this book is a report on the ongoing research of our technology transfer research group at the University of Texas along with colleagues with whom we are frequently in touch, if not in collaboration. Among these colleagues, some work directly in the process of technology transfer, while others research it. Many of our exchanges can be summarized around the following questions:

How do we identify or define technology transfer? Surely there are a large number of definitions that cover various aspects of this complex topic. In the centrifugal view we tend to see the broad process involved in bringing inventions to market. In the centripetal view, as we look into the details of this process, we often are concerned with communication phenomena—or, more typically, communication barriers. Many of the exchanges needed for successful technology transfer take place in situations not particularly opportune for effective communication. Another point that receives considerable attention is how to measure success. Obviously, from a market point of view, success is that the innovation got translated into profit. How do those involved in the details of the technology transfer process, however, gauge success? For example, a very useful instance of technology transfer communication might result in a valid understanding that an idea should not be pursued. Is this a "success"?

How can we organize so as to enhance technology transfer? Often discussion of this topic focuses on new organizational forms, such as R&D consortia, or other strategic alliances that have been created to promote technology transfer among organizations. Inherent in these forms is communication between central scientific units and shareholder companies, or among universities, industries, and government laboratories. One pervasive observation is also paradoxical: Although the explicit motivation for forming these new organizational units is to facilitate technology transfer, the communication problems associated with technology transfer actually tend to become more intense and concentrated within and across these organizations. A tradition of competitiveness does not yield easily to its more gentle counterpart—cooperation. It seems clear that organizations concerned with promoting technology transfer need to be flexible and adaptable to accommodate this shift.

What should be the roles of U.S. universities, state agencies, and industry as players in the technology transfer environment? Much attention usually is given to the contrasting environments of academia, federal laboratories, and industry in terms of value systems, timetables, communication styles, or reward systems, any of which can impede communication. It is recognized that many universities have a long tradition of discouraging applied research or involvement of faculty in outside activities, lest it distract from the university's missions in basic research and instruction. Universities have been cited for these shortcomings in the areas of providing continuing education and assisting with technology transfer. Campus-based innovations may be encouraged through formal transfer mechanisms, such as setting up centers for entrepreneurship, offering grants for research with a high transfer potential, and establishing reward systems for faculty who contribute to industrial development.

What is distinctive about the communication process of technology transfer? "Boundary spanning" seems to be an important concept for this issue. The communication involved in the technology transfer process often takes place between individuals using different vocabularies, styles, channels, schedules, and reward systems. Formal and informal communication barriers exist between different cultures, different sciences, and different levels of abstraction in what is being transferred (e.g., an abstract theoretical technology as against, say, a concrete product design). Serendipity may play a larger role in effective transfer than most people would like to admit; an important link may arise from a friendship, a crisis consultancy, or office or laboratory layout. Technology transfer does not necessarily lend itself to a rational mathematical model.

Toward a Communication Model of Technology Transfer

Our communication paradigm of source-destination (scientist-client, scientist-scientist, etc., as mentioned earlier) technology transfer draws in part from the earlier communication models formulated by such scholars as Wilbur Schramm (1971), David Berlo (1960), Marshall McLuhan (1964), and Everett Rogers (1983). Some of the contributions of coeditor David Gibson have appeared in previous works with Raymond Smilor and Everett M. Rogers in conjunction with their research in the Microelectronics and Computer Technology Corporation (Gibson & Rogers, 1990; Gibson, Rogers, & Wohlert, 1990; Gibson & Smilor, 1990; Smilor & Gibson, in press). In these earlier writings, as well as here, we consider technology as more than physical products; it is *information* that is put to use (Weick, in press). Technology transfer, then, is the iterative movement of this applied knowledge via one or more communication channels, with its communicating agents (scientists, clients, or "sources" and "destinations") being dyads structured as groups or organizations.

Given that the fundamental motive for communication among these agents is the transfer of the respective information, one can propose different allocations of responsibility among agents for the transfer to

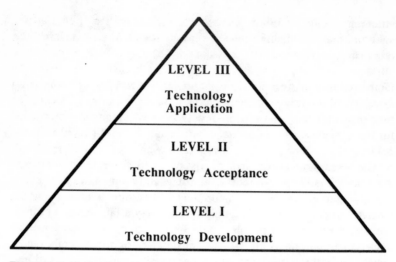

Figure A.1. Technology Transfer at Three Levels of Involvement

SOURCE: Gibson & Smilor (1990).

be successful. As shown in Figure A.1, the technology transfer phenomenon can be cast roughly into three levels which differentiate the goals involved in the transfer of a technology.

Level I: Technology transfer responsibility centers on the scientist who is conducting the state-of-the-art, precompetitive research, and who makes the results available to destinations (public or private) by a range of personal and mediated communications, such as technical reports and professional journals.

Level II: This is a level of shared responsibility between the source and its destination. The goal is to ensure that the technology is accepted, or at least understood, by someone who has the knowledge and the resources to apply and/or use the technology.

Level III: The most involved level of technology transfer includes shared responsibility for the application of the technology—namely, its profitable use, whether in the marketplace or by some other means.

A scientist's (person's, group's, organization's) control over the technology transfer process becomes increasingly problematic as one moves from Level I to Level III responsibilities, including the different measures of success at each level. Additionally, more cooperation is required of participants as one moves from Level I to Level III. To achieve timely and efficient Level III transfer, different functions, tasks, and networks must be activated simultaneously to overcome obstacles and barriers to the transfer process (Kozmetsky, 1988). To facilitate this goal, technology users (destinations) must be linked more directly and actively to the development of the technology. At the same time, researchers must gain a better understanding of and appreciation for the application of the technology (Gibson & Smilor, 1990; Smilor & Gibson, in press).

The foregoing levels also characterize some of the differences among prior attempts to model the technology transfer process. For example, the *appropriability model* (1945-1950s) emphasizes the importance of the quality of research and competitive market pressures to promote the use of research findings (Devine, James, & Adams, 1987). Such a model is consistent with a Level I responsibility pattern. Deliberate transfer mechanisms are viewed as unnecessary; all the researcher need do is develop the right idea (e.g., develop a better mousetrap) and the customer will beat a path to the inventor's door. Good technologies, according to this model, sell themselves.

The *dissemination model* (1960-1970s) emphasizes the diffusion of innovations where experts inform potential users of the technology (Rogers & Kincaid, 1981). This model has a Level II perspective. The objective is to transfer expert knowledge to the user who is a willing receptor. Once the linkages are established, the new technology will flow from the expert to the nonexpert much like water through a pipe once the channel is opened.

Much current literature on technology transfers reflects a *knowledge utilization model,* which emphasizes the importance of interpersonal communication between technology researchers and clients, and also identifies the organizational barriers or facilitators of technology transfer. Although this model begins to reflect the complexities of transfer from a Level III perspective, it suffers from an inherent linear bias (Dimancescu & Botkin, 1986). The stated or implicit notion is that basic research moves from researcher to client, in one direction, to become a developed idea and eventually a product. This model tends to reduce the transfer process to chronologically ordered one-way stages, whereas practice shows the process to be interactive and complex.

Several important characteristics of interpersonal communication underpin the *communication-based model* of technology transfer (Gibson, Rogers, & Wohlert, 1990). First, successful technology transfer is an ongoing, interactive process where individuals exchange ideas simultaneously and continuously. Feedback is so pervasive that the participants in the transfer process can be viewed as "transceivers," thereby blurring the distinction between the source(s) and destination(s).

The technology to be transferred often is not a fully formed idea and has no definitive meaning or value; meaning is in the minds of the participants. Researchers, developers, or users are likely to have different perceptions about the technology, which affects how they interpret the information. As a result, technology transfer is often a chaotic, disorderly process involving groups and individuals who may hold different views about the value and potential use of the technology. The communication-based model of technology transfer also can be seen as a particular case of the "garbage can model" of decision making proposed by March and Olsen (1976). Transferred technology is the result of an unplanned mixture of energy (participants), solutions looking for problems, choice opportunities, and problems looking for solutions. The model is not unidirectional. Feedback helps participants reach convergence about the important dimensions of the technology. Both problems looking for solutions (*technology pull*) and technology solutions looking for problems (*technology push*) are encountered.

About the Book

Because we believe the increasingly competitive global marketplace is the main stimulus for industrial and governmental interest in technology transfer, we have chosen to open this volume with a chapter by our colleague George Kozmetsky, a highly successful business entrepreneur in his own right. A former business school dean at the University of Texas at Austin and current director of the IC^2 Institute (also at the University of Texas at Austin), Dr. Kozmetsky has spoken and written for years on America's need for technological innovation and commercialization. In Chapter 1, "The Coming Economy," he describes technology as a key economic resource for competing in the global marketplace.

In Part II, we turn to detailed studies that attempt to examine the environments of technology transfer as it occurs both within and outside of organizational structures. These are chapters by our East Coast colleagues, Dorothy Leonard-Barton of Harvard University ("The Intraorganizational Environment: Point-to-Point Versus Diffusion") and

Ellen R. Auster of Columbia University ("The Interorganizational Environment: Network Theory, Methods, and Applications").

Specific contexts of technology transfer are the theme for Part III. Leading off this section is Christopher M. Avery's and Raymond W. Smilor's study of the nation's first major R&D consortium of the high-tech era ("Research Consortia: The Microelectronics and Computer Technology Corporation"). The importance of university-industry linkages is the topic of Chapters 5 and 6: "University and Industry Linkages: The Austin, Texas, Study," by G. Hutchinson Stewart and David V. Gibson, and "University and Microelectronics Industry: The Phoenix, Arizona, Study," by Rolf T. Wigand. Technology transfer also occurs in the context of businesses being spawned by federal laboratories, larger corporations, and universities, as described in Chapter 7, "New Business Ventures: The Spin-Out Process," by Glenn B. Dietrich and David V. Gibson. Finally, that the distance-cancelling uses of telecommunications networks can create a new type of "virtual" collaborative context is the topic of Chapter 8, "Transfer via Telecommunications: Networking Scientists and Industry," by Frederick Williams and Eloise Brackenridge.

The increasingly global character of the new economy, as seen in Part IV, is reflected in the interests of other nations in technology transfer as well as transfer across international boundaries. Examples of this international character are presented in Chapters 9 through 12 (Eduardo Barrera and Frederick Williams, "Mexico and the United States: The Maquiladora Industries"; James W. Dearing and Everett M. Rogers, "Japan: Tsukuba Science City"; Umberto Bozzo and David V. Gibson, "Italy: Tecnopolis Novus Ortus and the EEC"; and Arvid Singhal, Everett M. Rogers, Harmeet Sawhney, and David V. Gibson, "Bangalore: India's Emerging Technopolis"). This part is concluded with a chapter describing how technology transfer is accomplished in training programs for multinational corporations (Eun Young Kim, "Multinationals: Preparation for International Technology Transfer").

In Part V, we close the volume with a report on our preliminary bibliographic study of the literature of technology transfer. In "The State of the Field: A Bibliographic View of Technology Transfer," David V. Gibson, Frederick Williams, and Kathy Wohlert describe the topics and sources of current research literature. The chapter includes citations of literature we consider relevant to the study of technology transfer as a communication process, and which have not been cited in the book's previous chapters.

References

Berlo, D. (1960). *The process of communication*. New York: Holt, Rinehart, & Winston.

Devine, M. D., James, T. E., Jr., & Adams, I. T. (1987, fall). Government supported industry-research centers: Issues for successful technology transfer. *Journal of Technology Transfer, 12*(1), 27-28.

Dimancescu, D., & Botkin, J. (1986). *The new alliance: America's R&D consortia*. Cambridge, MA: Ballinger Publishing.

Gibson, D. V., & Rogers, E. (1990). *Texas high tech: Learning to cooperate for economic survival and growth*. Unpublished manuscript.

Gibson, D. V., Rogers, E., & Wohlert, K. (1990, June). *A communication-based model of technology transfer*. Paper presented at International Communication Association meeting, Dublin, Ireland.

Gibson, D. V., & Smilor, R. (1990). *Modeling key variables in technology transfer: Generalizing from the case of R&D consortia*. Unpublished manuscript.

Kozmetsky, G. (1988). Commercializing technology: The next steps. In G. R. Bopp (Ed.), *Federal lab technology: Issues and policies* (pp. 171-182). New York: Praeger.

March, J. G., & Olsen, J. P. (1976). *Ambiguity and choice in organizations*. Bergen, Norway: Universitetsforlaget.

McLuhan, M. (1964). *Understanding media*. New York: McGraw-Hill.

Reich, R. B. (1989, October). The quiet path to technological preeminence. *Scientific American, 261*(4), 41-47.

Rogers, E. M. (1983). *Diffusion of innovations*. New York: Free Press.

Rogers, E. M., & Kincaid, D. L. (1981). *Communication networks: A new paradigm for research*. New York: Free Press.

Smilor, R., & Gibson, D. V. (in press). Accelerating technology transfer in R&D consortia. *Research Technology Management*.

Schramm, W. (1971). How communication works. In W. Schramm & D. Roberts (Eds.), *The process and effects of mass communication*. Urbana: University of Illinois Press.

Weick, K. (in press). Technology as equivoque: Sense making in new technologies. In P. Goodman & L. Sproull (Eds.), *Technology and Organization*. San Francisco: Jossey-Bass.

I

Challenges of Technology Innovation and Transfer

1

The Coming Economy

GEORGE KOZMETSKY

*A new breeze is blowing across the world and is changing market
systems within and between countries. Government and business
leaders can no longer afford to passively regard the transfer of
technology and its commercialization. They certainly can no
longer take the technology transfer process for granted. There are
a large number of perplexing, often paradoxical and interrelated
drivers that are forcing the rethinking, reshaping, and restructuring of technology transfer for domestic economic well-being, for
global economic leadership, and for future harmony of the world
economy. In this chapter, noted business leader and former business school dean (as well as being an expert on innovation and
creativity in modern business) George Kozmetsky spells out a
strategy for change—a strategy where technology transfer is a
key component. Dr. Kozmetsky is Director of the IC^2 (Innovation,
Creativity, and Capital) Institute and Professor of Management
in the College and Graduate School of Business at the University
of Texas at Austin.*

The Importance of Technology Innovation

In the United States, the current budget deficits and trade imbalances, rising capital needs, more demanding community expectations for economic development, and especially more intense international competition in global markets now require leaders to develop better and newer ways to transfer and commercialize science and technology. For example:

> In this decade economic relations among industrial countries have fallen into critical imbalance, while stagnation has become the lot of many developing countries. The drift that typifies current policies, if not corrected, could lead to a disaster which none would escape. (Galbraith, Rostow, & Weintraub, 1988)

> Today, the concept of "technology exchange" is taking on a larger role. Those words now describe the process by which an entire nation harnesses its creativity and innovation on one realm—technology research—and translates that into leadership in a different realm: the competitive world of international business. (Rogers, 1988)

Technology innovation is taking on an ever-increasing importance for setting our future domestic economic course as well as establishing the nature of U.S. world economic leadership. Technology innovation is the process by which we go about harnessing our nation's research and development activities in science and technology and then transforming them into increased productivity for economic growth. It is the resulting economic growth that will help alleviate trade and budgetary deficits and economic problems facing this country as well as provide individual Americans with a higher real standard of living. Five important factors regarding the nature of technology and its commercialization in a modern economy have become clear.

- Technology is a constantly replenishable national resource;
- technology generates wealth, which in turn is the key to power—economic, social, and political;
- technology is a prime factor for domestic productivity and international competitiveness;
- technology is the driver for new alliances among academia, business, and government; and
- technology requires new managerial philosophy and practice.

The challenge of technology innovation in the coming economy can be stated as follows: How can we manage technology creatively and innovatively to reap the benefits of sustained economic growth? Those who do so will play the key roles in resolving the paradox of global competition and cooperation. What is the nature of this paradox? How do we find ourselves in it today?

Particularly since World War II, technological innovation—the entire process from R&D in the laboratory to successful commercialization in the marketplace—has been taken for granted. Traditionally, we have thought that successful commercialization of R&D was the result of an automatic process that began with scientific research and then moved to development, financing, manufacturing, and marketing. Managers and administrators continually strived for excellent performance in their respective institutions. They did not necessarily need to be concerned with linkages in the technology commercialization process. Insufficient attention was paid to the connections between the myriad institutions involved in moving scientific research from the laboratory to the marketplace. Most thought that there was little practical need to be concerned with the connections among academic, business, and government leaders involved in the commercialization process, in terms of understanding and evaluating their institutional performance for a better society.

Today's Setting

The recent pioneering 4-year study by the Office of Technology Assessment (1988) titled *Technology and the American Economic Transition: Choices for the Future* sets forth four major forces that are reshaping our society. These forces are:

- new technologies;
- loss of U.S. preeminence in international markets;
- the possibility that the price of energy and other natural resources may increase sharply by the turn of the century; and
- new values and changing tastes among consumers.

The report stresses that technology is a major driver for transforming the American economy in ways that are "likely to reshape virtually every product, every science, and every job in the United States (Office

Technology Assessment, 1988). The message of this important study that the American economy is at a crossroads and its future rests on set of conscious choices.

But do we really have a choice in pursuing leading-edge technolo-es? A good example is superconductivity. There is no question that movation in this technology will have an impact on a number of firms, idustries, products, and services, and will affect the security of our ation. It is a dual technology innovation having commercial and military applications. There is little choice but to pursue a technological movation with so many significant ramifications. Yet, according to an rticle on the Office of Technology Assessment draft report in *The Wall Street Journal* (June 20, 1988), "Japanese companies are poised to commercialize superconductivity technology well ahead of their U.S. rivals, despite the U.S. lead in basic research" (p. 2). More than choice is involved in the decision to pursue a technological innovation and its subsequent successful commercialization. Successful commercialization involves the ability to determine:

(1) when and how to take leadership positions in science and technological innovations and their commercialization;
(2) how to play catch up when others have manufacturing and market leadership; and
(3) how to leapfrog competition to become pertinent both scientifically and economically.

A number of leaders in government, academia, and business have made it clear that past policies, theories, doctrines, and practices are no longer adequate to address these types of choices wisely. Peter F. Drucker (1986) forcefully stated that the world economy already has changed irreversibly in its foundations and structures: "It may be a long time before economic theorists accept that there have been fundamental changes, and longer still before they adapt theories to account for them" (p. 768). The noted scholar on technological innovation Dr. Ralph Landau (1988) pointed out that technological change has been central to U.S. economic growth, both directly (in which case it can be said to account for perhaps 30% of economic growth) and through its positive effect on other factors of production (which can be said to account for perhaps another 40% or 50% of economic growth). Although scholarly research has not provided conclusive evidence yet, it is in this broad sense that technological innovation is the key to viable strategies for future economic growth: It can raise the factor productivity of the economy at an accelerated rate (Landau, 1988).

Today, there is a major transformation underway in which all leaders need to rethink technology's role in economic change domestically and internationally. What does it mean to today's leadership to take viable strategies for stable economic growth? Research at the IC^2 Institute of the University of Texas at Austin shows that maintaining global market competitiveness is requiring more and more heavy capital commitments to new, more risky, often large-scale technologies that require 5 to 10 years to implement successfully. Often these choices are "bet the company" or "bet a whole American-based industry" decisions. At the same time, the key to this country's great economic gains has been the speed with which bright innovators have recognized technology potentials and then used their know-how to exploit them. From a leadership perspective, rethinking technology innovation's role in economic change requires a restructuring of the way

(1) science and technology are developed and transferred;
(2) businesses are managed; and
(3) rules, regulations and incentives operate within the private and public sectors to use technology innovation in a timely manner.

Technology has provided a range of opportunities and challenges to all of our leaders in academia, business, and government. When viewed from an international perspective, the challenge to speed up our nation's rethinking process becomes more imperative. Why is this so? Technology has reshaped the products, services, and jobs in all nations, not just the United States. The Japanese story is well known. Gorbachev's (1987) book *Perestroika* sets forth what he calls "the thinking for USSR and the world." In terms of technology, and the USSR's "goal to meet world technological standards," Gorbachev states:

> We decided to put a firm end to the "import scourge" as our economic executives call it. To these ends we are putting into operation the great potential of our science and mechanical engineering. . . .

> We are launching target-oriented programs, prompting work collectives and economic and other scientists to work in a creative way, and have organized twenty-two intersectoral research and technological complexes headed by leading scientists. . . .

[The CPSU Central Committee] set a target unprecedented in the history of the Soviet industry, that of reaching in the next six to seven years world standards as regards major machinery, equipment, and instruments. (pp. 94-95)

The setting for rethinking technology's role in shaping the global economy can best be described as a lack of consensus "rules of the game." Of course, there are well-known rules for patents, trading practices, and so forth; but there is no discipline forcing countries to pursue more compatible policies. Advances in technology—regardless of where they are made or who makes them—will be reflected in national economic structural changes that will compound the problems of macroeconomic change in international trade and payments. This means that leaders, while they are making domestic choices, will need to take into account their impact on other nations' economic structures. For example, Japan has used its technology to make significant inroads into the world's semiconductor markets while maintaining a competitive national market.

The USSR and the Japanese examples make it clear that it is necessary to differentiate international competition from domestic market competition. International competitiveness reflects differences in ways technology is developed, transferred, and commercialized. The more specific differences lie in the relationships between each national government and its private sectors and the cultural differences that affect how managers measure success in commercializing technology innovations (e.g., return on investments, profits, employment stability, market share, short-term gains, long-term benefits, and relative change in productivity). In the future world economy, these differences can well change the world's economic leadership that can, in turn, change regional relationships for extended periods of time.

How do we then put our arms around the changes, paradoxes, issues, crises, changes, concerns, and problems of today's setting for understanding the role of technology innovation in the coming global economy?

Any discussion of the economic future must have a clear notion of how our society "hangs together"—how its parts are related to one another, and which elements are more susceptible to change than others. These discussions also must take into account the interplays of values, motivations, and resources. Leadership needs to provide the necessary catalyst that assures a creative environment for change which empowers others to translate intentions into realities and sustain them. Our country needs more than managers who do things right; we need leaders who do the right things.

Understanding the Technology Commercialization Process

Commercialization is the process by which R&D results are transformed into the marketplace as products and services in a timely manner. It requires the active interchange of ideas and opinions that are both technological and market-oriented in nature. The commercialization process may benefit through increased scale of production, higher quality, and lower prices. Commercialization helps to define the educational and training requirements for present and emerging marketplaces. It can also be a major driving force that invigorates emerging industries and rejuvenates older industries.

The traditional and still-current strategy in our nation for commercializing technology is that industrial laboratories concentrate on mission-oriented projects, and that universities confine themselves primarily to basic research and teaching. During the past decade, advocates for changing this strategy recognized "gaps"—namely, the loss of global markets, hollowing out of our manufacturing plants and piecemeal collaborative efforts for research and development, joint ventures at home and abroad, the buying out of American firms, and the growth of foreign companies in the United States. The use of technology as a resource traditionally has been perceived as an individual institution's choice and responsibility. Economic developments would flow from this process because of American ingenuity and our entrepreneurial spirit. It was expected that all regions of the United States would in time enjoy the benefits of this paradigm in which new innovations from research were followed naturally by timely developments, commercialization, and diffusion.

This traditional paradigm—that science and technology naturally evolve into commercialization—has become clearly inadequate. In the emerging economy, it does not adequately (1) provide sufficient employment opportunities; (2) mitigate layoffs and plant closures; (3) maintain a strong, global, and comprehensive security position; (4) provide regional and local economic development; and (5) present growth opportunities across the board for basic and high technology industries.

Since 1976, a new paradigm has been emerging to meet the afore-mentioned realities. It includes institutional developments involving academia, business, and government technology venturing. These developments seek to maintain our leadership in science and technology by accelerating the successful commercialization of innovation in a competitive environment. The underlying assumption in this paradigm for commercialization is that scientific and technology innovation leadership can result in industrial and product leadership if our scientific talents are concentrated on applications selected by those who could commercialize them.

Throughout America, most of the states are moving to aid the growth of high technology and to foster technological commercialization. They, rather than the federal government, have taken leadership roles through policy development organizations, economic growth initiatives, and corporate-university partnerships. Since 1980, hundreds of programs or initiatives have been developed by states for high technology and economic development. For most states, the major source of R&D funds is still the federal government. There is a direct correlation between those states receiving the largest federal obligations for R&D and those taking the lead to initiate technology-venturing developments.

As a result, these activities are bringing together in dynamic and interactive ways state governments, local governments, private corporations, universities, nonprofit foundations, and other organizations. They are developing corridors and triangles between key cities or research universities. Centers of excellence are appearing within these corridors and triangles. They have begun to lay out science and research parks and target emerging industries for long-term industrial growth. Leadership networks are forming between previously isolated institutions. These developments are not accidental. These new institutional developments are managed efforts for economic growth and diversification that (1) develop emerging industries; (2) provide seed capital for early and start-up entrepreneurial endeavors; and (3) assure U.S. scientific/economic preeminence.

Developing Emerging Industries

Institutional relationships involved here are academic and industrial collaborations and industrial R&D consortia. Because the basic research is carried out in universities and colleges, collaborative efforts between academia and industry can accelerate the commercialization of basic research into emerging industries.

Since 1982, there have been a series of private corporation joint-research efforts. The obstacle to such consortia in the past has been ambiguous legal status under the antitrust laws that would entail the risk of huge penalties. The passage of the Cooperative Research and Development Act in October 1984 has done much to alleviate these legal concerns. As a result of the passage of this act, there has been a proliferation of large-scale R&D consortia, which now number more than 150 nationwide.

Some forms of institutional development such as incubators, SBIR programs, state venture capital funds, and risk capital networks are providing seed capital for small and take-off companies. They are also pushing regional and local economic diversification through entrepreneurial activities.

Some states now are developing special initiatives such as tax incentives, enterprise zones, research and industrial parks, and direct financial assistance in the form of low-interest bank loans and loan guarantees to meet the start-up capital needs of new technology firms.

Assuring U.S. Scientific/Economic Preeminence

A number of institutional developments are seeking to ensure U.S. scientific and economic preeminence. The focus on the creation of National Science Foundation (NSF) centers of research and engineering excellence, government/business/university collaborative arrangements in technological areas, industrial R&D joint ventures and consortia, and NSF's sponsorship of Industry/University Cooperative Research Centers (IUCR). These are intended to provide a broad-based research program that is too large for any one company to undertake alone.

Scholars and practitioners are reassessing the entire nature and direction of technology innovation and its commercialization. They are raising important questions about its effectiveness, theory, and direction. As a result, we find ourselves in a diverse and lively muddling process, trying to better understand the key factors involved in technology commercialization. On a variety of fronts, there are challenges on how we organize for and take advantage of technological innovation in a period of hypercompetition. For example:

- "Intellectually, there is a major problem to be sorted out. There is a powerful historical tendency for late comers to catch up with early comers." (W. W. Rostow, personal communication, 1988.)

- "The ideal of the entrepreneur has been taken to excess . . . the constant spawning of new companies may be actually sapping America's economic might." (Pollack, 1988, p. 1.)

- Consortia members "will then 'compete away' part of the returns . . . To that extent, none of the members will benefit." (National Science Foundation, 1988, p. 2.)

- Setting priorities on science has been "a subject heretofore considered inappropriate for mention in polite scientific society. The question remains as to how this is to be done." (*Science* editors, May 20, 1988, p. 965.)

- "Our report makes one principal recommendation: Universities and corporations need to assess the results of their collaborative research and technology exchange efforts more systematically and effectively. This recommendation is offered in the conviction that the results of such assessments will further simulate productive cooperation." (*Beyond the Rhetoric,* a report by the Business-Higher Education Forum, May, 1988, p. 9.)

- "A strong science base supplies a vast storehouse of new ideas, and a good educational system provides engineers and manufacturing workers with knowledge; but strength here cannot make up for inadequacies in the functioning of the development and manufacturing cycle." (Gomory & Schmitt, 1988, p. 1204.)

It is apparent that we lack the necessary integration for successful technology commercialization. The requirements for developing such an integration in the coming economy demand the simultaneous linking of several factors:

(1) *need*—Is there motivation for pride, security, and profit?

(2) *vision*—Are there adequate mentors to foster creative collaborative programs?

(3) *timing*—Is there a real current market need?

(4) *technical feasibility*—Does the technology exist? Is the required research underway?

(5) *skilled personnel*—Is there sufficient quality and quantity?

(6) *champions*—Is leadership available and in place?

(7) *adequate financial resources*—Has the necessary capital been committed?

(8) *public and political support*—Is there a consensus on key public policy issues? How will it be sustained?

(9) *public/private sector cooperation*—Are the institutional alliances viable and effective?

(10) *intellectual community support*—Is the theoretical basis being developed in harmony with the new reality?

Successful technology commercialization usually takes longer and costs more than anticipated, and failures surely will occur along the way. Consequently, there must be contingency planning to incorporate the following needs:

(1) continuity of public and private support;

(2) capital infusion;

(3) parallel developments to alternative approaches and supporting technology;

(4) more skilled personnel than originally planned;

(5) the need for three types of leaders: one to start, another to move forward, and still another to take it to the marketplace.

The relationship between technology innovation and economic wealth generation, resource development, markets, and job creation involves more than making economical investments. Just as important as the usual economic standards of effectiveness and efficiency are two other dimensions. First, flexibility: If we do not make the investments, will we still have sufficient flexibility to meet competitive challenges? Second, adaptability: If we do not make these investments, can we adapt ourselves to unexpected or unforeseen competition without serious consequences to our firms and communities? These are the challenges—flexibility and adaptability—that technology innovation poses.

Rethinking, Reshaping and Restructuring for the Coming Economy

Meeting these challenges requires that we rethink, reshape, and restructure our approach to technology innovation for the coming economy. Rethinking will lead to a better definition of the technology innovation process which, in turn, will influence national policy, industry development, and firm activity. Reshaping will lead to a more

effective model of the economy that reflects the importance of networking and will help delineate alternative approaches to technology innovation. Restructuring will form the foundation for the new industries
emerging from leading-edge technologies.

Rethinking

The technology innovation process requires a kind of parallel processing; many things have to be going on at the same time at many
places. Technology innovation is not a clear-cut, step-by-step process.
Linking R&D to the marketplace seems to most people to be a chaotic,
complex, convoluted, and very messy process. But there is a connective
pattern to this chaos. The pattern is a parallel one; actions need to be
taken simultaneously in myriad institutions to move the technology to
the market continually.

There is also a serious process at work at each nodal point of the
movement of a technology from the laboratory to the marketplace.
Some matters have to be done serially, such as individual projects. But
each of these serial operations can, in many cases, be going on simultaneously. To illustrate this point, three examples follow: one at the
national policy level; another at an industry level; and a third at a
company level.

At the national level, the Japanese approach is interesting. It promotes many functions—like the following six, all of which have serial
components but all of which are carried on in parallel:

1. International research cooperation
   a. specific international joint research projects
   b. summit projects (e.g., Human Frontier Science Program)
   c. invitation of foreign researchers to Japan
   d. collaboration with developing countries in joint research
      projects
2. Technological development programs
   a. R&D basic technologies for future industries
   b. large-scale projects
   c. R&D project for medical and welfare equipment
   d. energy-related technologies
3. Regional economics
   a. R&D on important regional technologies
   b. technopoleis[1] program
   c. national research laboratories

      d. joint government-private sector research
4. Promote development of technology in the private sector
    a. Japan Key Technology Center
    b. conditional loans for development of technologies to activate industry in basic materials, and core and priority technologies; and to innovative basic technologies, environmental-protection safety technologies, regional technologies, and energy-conservation, power-generation, and oil-substitute technologies
5. Promotion of industrial standardization
    a. voluntary standards for quality assurance and interchangeability of products for both domestic and foreign manufacturers
    b. research for standards for material, textile and chemicals, machinery, electrical and electronic, and information standards
    c. participation in international standardization
6. Diffusion of technology accomplishments
    a. patents and licensing of state-owned intellectual properties
    b. international exchanges of technology

At an industry level, Debra Rogers has described the parallel approach as "the process by which an entire nation harnesses its creativity and innovation in one realm—technology research—and translates that into leadership in a different realm—the competitive world of international business" (Rogers, 1988, p. 1). Her technology transfer continuum is an effective representation of this parallelism with its multiple feedback processes. She further describes the infrastructure for commercialization of technology at an industry level in terms of structure, resources, and methods and tools available. The process itself is parallel. It is stimulated by creativity and implemented successfully through innovation with parallel processing being coordinated through three institutions: education, government, and industry (see Figure 1.1.).

At a firm or company level, IBM is concerned with commercializing technology innovation as a two-step process with strong connections between development and manufacturing cycles. The first step is when a creative idea or technology dominates and is innovatively implemented in a specific product. The second step involves cyclic development or repeated incremental improvement to the product engineered by the first process. The repeated incremental advances result in product improvement with new features year after year. Although the process is evolutionary, the cumulative effects are profound in terms of market domination. As R. E. Gomory and R. W. Schmitt (1988) have pointed out: "In technology areas where the United States has not been

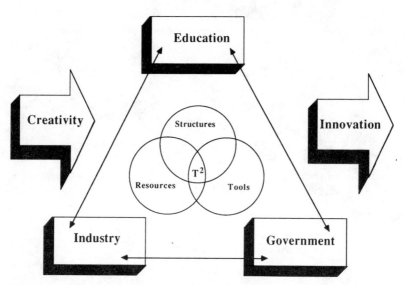

Figure 1.1. The Planning Model: Technology Transfer Elements
SOURCE: Rogers (1988).

competitive, we have lost, usually not to radical new technology, but to better refinements, better manufacturing technology, or better quality in an existing product." Consequently, it is important to develop a parallel process within the firm. Then, when a product is in the manu- facturing stage, it is possible to have the development team working on the next creative idea or technology.

Professor Laura Tyson of the University of California, Berkeley, was correct when she stated, "Simply put, in the long run, investment in science, education, technological innovation and technology diffusion is all that will sustain us" (Starr, 1988, p. 120). The challenge we face in making these investments is how we balance individual choices as investors, voters, consumers, and employers with institutional choices that affect the nature and scope of our nation's economic future. Re- shaping our economy based on individual choices and connecting them to the requisite investments in science and technology is a difficult task. It is confounded further when we try to link these investments and their utilization with the myriad of institutions involved.

The American emphasis has been on individual capabilities. In this sense Americans excel, according to Professor Iwao Nakatani (1988), professor of economics at Osaka University:

> A close look at those industries in which U.S. firms are competitive reveals that most belong to fields which attach little importance to coordination between people and offices, with success or failure usually hinging on the effective use of individual capabilities. This category would include such activities as basic research at universities and research institutes, the design of systems and software for computers, research and development of pharmaceuticals and aerospace industries such as missiles and aircraft. In these fields, individual creativity is crucial, while coordination does not hold much weight. These two factors are essential conditions for any industry in which the United States is to remain competitive.
>
> In assembly and processing industries, however, a wide variety of parts are brought together from numerous suppliers and assembled through precise operations. This requires close links between people and organizations, as well as careful coordination of activities. Ordinary, day-to-day innovations generated in these relationships and interactions thus determine the competitiveness of the industry. These are precisely the industries in which the U.S. is having the most difficulties.
>
> Overall, it could be concluded that the U.S. will hold onto its competitive edge in those industries . . . in which individual ability is the prime criteria for success. In industrial sectors which stress coordination, however, it will be difficult for the United States to regain competitiveness even if the dollar plunges further.
>
> It is imperative that the United States and Japan join hands to complement each other's weak points. While the strength of Japanese firms in terms of coordination is beyond question, they cannot even come close to matching their U.S. counterparts in making effective use of individual capabilities. U.S. corporations, for their part, generally experience great difficulty with "horizontal relationships," and could perhaps learn a great deal from the patience which characterizes Japanese corporate management. What is required of U.S. and Japanese corporations is not mutual hostility, but rather, mutual cooperation.

Reshaping

Reshaping of the American economy requires a framework to help chart its future direction—a direction that is being influenced by the result of myriad individual private choices, by the way businesses are being managed by rules, regulations, and incentives being adopted by the public sector, and by the use of emerging technologies. Since the first industrial revolution, technology innovation has been a basic motor for economic growth and job creation. For example, much of our economic growth in the past decade and a half has been influenced by information-technology innovations, coupled with job creation through small business firms.

As Table 1.1 shows, by 1983 more than 50% of the U.S. gross national product (GNP) was derived from information technologies. Small business establishments generated 62% of the net employment increase between 1974 and 1984 (U.S. Small Business Administration, Office of Advocacy, Small Business Data Base.)

Reshaping our approach to technology innovation must reflect the new realities of internationalization, interdependence, and networking. The OTA study (Office of Technology Assessment, 1988) has set forth basic characteristics of the future American economy that effectively reflect these important themes:

(1) Each part of the economy is interdependent with the other, forming a conceptual network.

(2) Technology can increase the efficiency of the whole network and each part of the network.

(3) Technology can increase greatly the efficient use of energy and materials if investments are made for design of the parts and their management.

(4) Problems in one part of the economy quickly impact on many other parts both directly and indirectly.

(5) New connections grow quickly between parts of the economy.

(6) Many parts of the economy have become internationalized and move about the world with ease and speed such as products, finance, and ideas.

(7) As the connections become more complex and as things move in parallel, the role of transactional business people and professionals grows as well as their costs.

(8) Businesses are locating where they can find adequately skilled and trained personnel in sufficient quantity and where extensive networks of personal contact can be maintained.

(9) Technology has changed the nature of work done at home and at work.

(10) Successful adaptation of the new economy will require skills provided by a solid basic education.

TABLE 1.1: Character of Output, U.S. Informational Economy Revenue Estimate, 1958 and 1983 in Billions of Dollars

	1958[1]	1983[2]
Education	$60.2	$218.7
Research and Development	$11.0	$87.0
Media and Education	$38.4	$244.8
Information Services	$18.0	$967.5
Information Machines, Manufacturing	$8.9	$169.3
Total	$136.4	$1,687.3
GNP	$475.6	$3,304.8
% of GNP	28.7	51.1

SOURCES: 1. Machlup, F. (1962). *The production and distribution of knowledge in the U.S.* (pp. 354-357). Princeton, NJ: Princeton University Press.
2. IC[2] Institute. (1968). *Commercializing technology resources for competitive advantage* (p. 4). Austin: University of Texas.

This report calls further attention to the following:

> There is a paradox in all of this. Countries, establishments, communities, and individuals are finding themselves more tightly connected. And yet, the networks allow more independence and choice. In particular, technology may tie production systems in different countries more closely together while nations become less and not more dependent on imported supplies of energy, food and manufactured products. In such situations, the movement of materials may have decreased while the strength of linkages moving information, technology, ideas, and capital equipment has increased.

As a result, we must reshape our use of technology innovation. How can we begin to do this? Among alternative approaches are the following:

(1) Continue to have study commissions with reported recommendations, and hold symposia and conferences on the topic of technology innovation and its commercialization.

(2) Foster and speed up better research to understand more fully the impacts of technology innovation and to forecast their changes on emerging industries and economic growth.

(3) Hold a national congress on technology innovation with a focus on its parallel-processing dimension.

(4) Hold national conferences on technology entrepreneurship to discuss incentives and barriers to technology innovation in small companies.

(5) Convene an international commission to provide direction on the world economy and world policy.

(6) Keep the status quo, thus letting technology evolve and basing reactions on production needs, individual choices, and crises.

The timing is ripe for reshaping our approach to technology innovation. The world political leadership is undergoing rapid change. The USSR, China, and Japan very well could have new leaders, the Western European nations are preparing busily for new leadership in 1992, and Eastern Europe is changing daily. The next few years provide an unprecedented opportunity for the world community to address the issues, concerns, and paradoxes that surround technology innovation and commercialization. Particular needs to be addressed include reciprocity in technological exchanges; intellectual property rights; early access to new products, processes, and equipment; foreign investments in other nations; and work force requirements.

Restructuring

Emerging industries are those that will be developed by using the revolutionary technologies that only now are beginning to move from invention to innovation, such as lasers, new materials, biotechnology, international telecommunication networks, very large integrated circuits, medical instrumentation, superconductors, and breakthrough managerial methodologies. These technologies have the following four characteristics, as pointed out by W. W. Rostow (1986, 1988):

- They are so encompassing that no one country can dominate them completely.
- They are linked to areas of the basic sciences that also are undergoing revolutionary changes.
- They are immediately transferable to rapidly industrializing nations.
- They are key to leapfrogging for basic industries.

The emerging industries based on these technologies are being challenged across all dimensions, from scientific development to invention, to innovation, and to commercialization. These industries must rely on government and academia to help meet the scientific and technical challenges upon which their viability depends.

If we are to maintain U.S. preeminence in these emerging industries, then we must find more effective ways to maintain a global competitive edge in these arenas. Maintaining political and economic manufacturing advantages requires that we begin to think and act in new ways to commercialize technological resources, wherever they may be.

If the United States is to meet the competitiveness challenge facing this country, then we must take more of a global view. It is necessary to recognize that there is more and more emphasis on locating the best possible places in both economic and political terms around the world to conduct R&D, manufacturing, marketing, and financial activities. There is also more and more emphasis on the world as a single marketplace, rather than one that is subdivided nationally or regionally. Yet, it is best to describe natural reactions to this today as protectionist, intensely competitive, and at best muddled.

Global competitiveness focuses on developing global relationships for science and implementing cooperative and competitive strategies among trade partners and allies. Competitiveness and cooperation seem to be a paradox; but they actually are the cornerstones for a more effective U.S. strategy for commercialization.

Competition is the outcome of those national attributes that help individuals and firms to perform more effectively and efficiently. Consequently, it enhances the relative strength of companies, facilitates international trade, and adds to the world's quality of well-being. Cooperation among scientific and technological organizations can become an important source for emerging industries. Improved processes for commercialization, for example, require more effective interaction among federal labs, industrial labs, and universities.

Technological innovation in the coming economy demands an integrated, holistic approach that blends technological, managerial, scientific, socioeconomic, cultural, and political ramifications in an atmosphere of extreme time compression. This approach centers on new ways for government, business, labor, and academia to work together.

In all, the United States still has the world's most creative technology base, a stable political system, a world-class higher educational system for research and teaching, the world's largest market, and a large capital base. We have a tradition of entrepreneurship and a demonstrated ability to respond rapidly to severe crises. We have the resources and the know-how to compete globally, to cooperate with other key nations such as Japan in providing world economic leadership, and to meet the challenge of technological innovation in the coming economy. We need to make the best use of these resources in our preparation to compete in the coming economy.

Note

1. The plural form of the Greek word *polis* is *poleis*; we have chosen in this book to use *technopoleis* as the plural form of *technopolis* rather than the awkward-sounding *technopolises*.

References

Drucker, P. F. (1986). The changed world economy. *Foreign Affairs, 64*(4).

Galbraith, J. K., Rostow, W. W., & Weintraub, S. (1988). *Proposal for a high level report by an international commission on the future of the world economy (World Maekawa Report)*. The University of Texas at Austin.

Gomory, R. E., & Schmitt, R. W. (1988, May 27). Science and product. *Science, 240.*

Gorbachev, M. (1987). *Perestroika: New thinking for our country and the world.* New York: Harper & Row.

Landau, R. (1988, June). U.S. economic growth. *Scientific American, 258*(6).

Nakatani, I. (1988, June 11). Japan, U.S. industry must meld strengths for mutual benefit. *The Japan Economic Journal.*

National Science Foundation. (1988, March). *The role of the national science foundation in economic competitiveness*. National Science Board Committee, draft final report. Washington, DC: National Science Foundation.

Office of Technology Assessment. (1988). *Technology and the American economic transition: Choices for the future*. OTA-TET 283. Washington, DC: Government Printing Office.

Pollack, A. (1988, June 14). High-tech entrepreneurs: New doubt on a U.S. ideal. *New York Times*, p. 1.

Rogers, D. M. A. (1988, February) *Toward a national campaign for competitive technology transfer.* Paper presented at the 1988 AAAS Annual Meeting, Boston.

Rostow, W. W. (1986, September-October). Economic growth and the diffusion of power. *Challenge.*

Rostow, W. W. (1988). The fourth industrial revolution and American society: Some reflections on the past for the future. In A. Furino (Ed.), *Cooperation and competition in the global economy: Issues and strategies*. Cambridge, MA: Ballinger.

Starr, M. K. (Ed.). (1988). *Global competitiveness: Getting the U.S. back on track*. New York: Norton.

II

The Organizational Setting of Technology Transfer

┌─────────────┐
│   │
│   2   │
│   │
└─────────────┘

The Intraorganizational Environment: Point-to-Point Versus Diffusion

DOROTHY LEONARD-BARTON

Technology transfer is difficult both to research and to manage—in part because the circumstances surrounding the interaction between technology development sources and technology receivers differ from transfer to transfer, even within the boundaries of a single organization. The circumstances, however, are not totally idiosyncratic. That is, although there is no one universally applicable model, two distinct transfer situations prevail in practice and in academic literature: point-to-point and diffusion. Using these two situations to identify ends of a continuum, this chapter explores the differing characteristics of each that lead to variance in management practices. Illustrative evidence is drawn from 51 case studies of one particular kind of transfer, namely, the invention and dissemination of productivity- or production-enhancing tools within for-profit corporations.[1] Dorothy Leonard-Barton is an Associate Professor at the Harvard Business School.

AUTHOR'S NOTE: The research reported here was supported by the Division of Research, Harvard Business School. An uncut version of this chapter, which includes examples, may be obtained by writing the author.

Distinguishing Between Point-to-Point
and Diffusion Technology Transfer Situations

The two modes of technology transfer—point-to-point and diffusion—have been separated in much of the academic literature more because of the authors' disciplinary backgrounds than because of any conscious conceptual distinctions (exceptions being Bradbury, Jervis, Johnston, & Pearson, 1978). Thus, researchers writing about the commercialization of new technologies have tended to feature point-to-point transfer—that is, transfer from a single source to one receiver site—(e.g., see Rubenstein, 1989; Souder, 1987)—as have economists discussing international technology transfer, using the country as the unit of analysis (e.g., Mansfield et al., 1982). In contrast, sociologists, geographers, and marketers concentrating on the behavior of large groups of individuals have written about diffusion, that is, transfer from a single source to multiple receiver sites (see Brown, 1981; Midgley, 1977; Rogers, 1983). On those occasions when academic literature or the trade literature discusses the two modes simultaneously, the discussion often mirrors common practice in the field by ignoring differences between the two.

Distinguishing between technology transfer situations is useful for both practitioners and academics. First, implementation managers often do not recognize that there are two different transfer situations, because both modes involve the same basic process—namely, the initiation and implementation of a new technology—and because in both situations technical innovations are deployed within organizations. Yet it is important to recognize, from the beginning of a technology development project, what kind of transfer one is involved with in order to (1) establish user expectations; (2) allocate resources appropriately; and (3) choose transfer managers with the right skills. It can be inefficient or even risky for technology developers and implementation managers to misunderstand what kind of transfer situation they are managing. Users of new technology similarly need to understand through which transfer modes they are receiving, in order to allocate resources appropriately.

Explication of technology transfer modes is relevant to academics, because independent variables such as transfer tactics (especially user involvement) and appropriate managerial strategies differ among transfer modes. Empirical findings reported in the literature therefore need to be considered within the context of the particular transfer situation providing data, if those findings are to fulfill the promise of social science—which is the accretion of cumulative knowledge.

Defining Technology Transfer

In academic literature, the term *technology* has been applied widely to everything from manufacturing hardware (Woodward, 1965) to search procedures (Perrow, 1967) or skills possessed by people (Rousseau & Cooke, 1984).[2] The definition most applicable to this chapter is Donald Schon's (1967): "any tool or technique, any product or process, any physical equipment or method of doing or making, by which human capability is extended." Thus, *technology* is capability, that is, physical structure or knowledge embodied in an artifact (software, hardware, or methodology) that aids in accomplishing some task. These definitions deliberately exclude from discussion expertise that is "found inside one's head" (Ulrich & Weiland, 1980, p. 87) or "performance programs stored in individuals' memories" (Gerwin, 1981, p. 5). Such knowledge is defined here as technology only when it is captured at least partially in some communicable form.

The term *transfer,* as used here, refers to transformation of a technical concept of proven feasibility into a development state closer to its end use in the production of a service or goods (including, possibly, another tool or machine). To the extent that the transformation is complete and receivers find that the new tool enhances their process or output in some way, the transfer is successful. *Technology transfer* as thus defined usually involves some source of technology, possessed of specialized technical skills, which transfers the technology to a target group of receivers who do not possess those specialized skills and who therefore cannot create the tool themselves.

Within organizations, the development of technical concepts into usable tools often follows the progression shown in Figure 2.1. Transfer, even of the same technology, can occur at multiple points in this progression. The research organization passes the technology to the development organization, which passes a revised version on to engineering (with iterations possible at each point of passage). Thus, the same group of people may be technology sources in one instance and receivers in another. This chapter focuses on the latter two linkages and mostly linkages within a firm, rather than those with outside sources.[3]

1) Basic Research:
Exploration of basic
scientific principles
↓

2) Applied Research (Development):
Focus of scientific principles
on specific applications
↓

3) Engineering:
Development of the applied
principles into a
product/tool
↓

4) Implementation (Use/Manufacture):
Use of that tool or product to
accomplish a task/production
of the product in volume

Figure 2.1. Stages of Technology Development within a Corporation

Defining Technology Transfer Modes

The key dimension distinguishing the two ends of the technology transfer mode continuum is the number of individuals targeted as users for a particular application of the technology. The fewer the user/receivers per technology application, the closer the situation is to a pure point-to-point transfer, the theoretical extreme of which would be a custom-made tool for one user. At the other extreme would be diffusion of a generic tool to thousands of users.[4]

Each mode is complicated by the degree of diversity in the applications of the tool, that is, the number of different tasks the tool is to aid. At the low end of this continuum, all users apply the tool to the same task; at the high end, users with diverse jobs use the same tool but apply it to very different tasks. These two dimensions, innovation span (number of people) and innovation scope (number of different applications), combine to create four technology transfer situations: *simple, complex,* point-to-point, or diffusion.[5] Before refining the discussion by exploring all four quadrants in Figure 2.2, however, we will first examine the

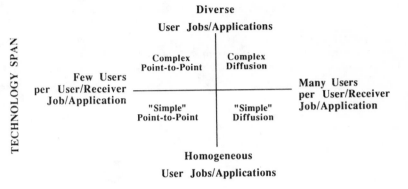

Figure 2.2. Four Modes of Technology Transfer

differences between the simple forms of point-to-point and diffusion (the horizontal axis). The volume and physical dispersion of users determine the kind of user representation possible during the development of the new tool, the degree to which the tool can be customized, and the problems that implementation managers will be most likely to encounter (see Table 2.1).[6]

Point-to-Point

When a single group of technical experts transfers the technology to another single targeted group, the mode is point-to-point. The first two handoffs presented in Figure 2.1 (between research and development, and from development to engineering) are almost always point-to-point.

Once the technology is embodied in a tool or product by some internal corporate group performing the development/engineering function, however, it may be deployed into use either through a point-to-point transfer or through a diffusion campaign. Tools that involve a large capital investment and that are used by a few people in some central, critical step in a process stream are likely to be deployed bilaterally through a point-to-point transfer.

Process tools that are intended for a large number of sites or users—or both—within a large organization, such as standard computer-aided

TABLE 2.1: Characteristics of Two Primary Modes of Technology Transfer

	CHARACTERISTICS OF MODES	
	Point-to-Point	Diffusion
DESCRIPTORS:		
1. Organizational Span	Narrow	Wide
2. Geographic Dispersion	Narrow	Wide
MANAGERIAL ISSUES:		
1. User Representation in Design	Actual	Representative or Virtual
2. Customization	Single, Highly Customized Application	Limited, Standardized Application
3. Communication Mode	Negotiation	Marketing

design tools for engineers or test tools to be used in manufacturing, are more likely to be disseminated through a process of diffusion.

Diffusion

Diffusion of innovations, the study of "the process by which an innovation is communicated through certain channels over time among the members of a social system" (Rogers, 1983, p. 5), is a very well-known model of technology transfer, both among sociologists and market researchers. It sometimes is compared to contagion (Brown, 1981) because of the importance of the horizontal (receiver to receiver) spread after the first, vertical dissemination (technical source to receiver) (Leonard-Barton & Rogers, 1981).

The diffusion model has been applied much more frequently to the spread of innovations among individuals in the marketplace, outside of an organizational setting (Gatignon & Robertson, 1985), or within an industry with each corporation considered as a single adopter (Czepiel, 1974; Mansfield et al., 1977), than it has to the spread of innovations within a corporation. The reason for this emphasis is that "diffusion researchers have been more oriented to the dependent variable of adoption (a decision to use and implement a new idea), than to actual implementation itself" (Rogers, 1983, p. xvii). Therefore, scholarly investigation usually has stopped at the point at which some senior manager—or group of managers representing the whole organization—has decided to adopt the innovation, but has not proceeded to investigate its subsequent diffusion throughout the organization.

Within the social system of an organization, there is a diffusion process that conforms in many ways to the sociological model as defined by Rogers (1983). For instance, communication about the innovation is important in stimulating or hindering diffusion. Just as within any other social system, there are early adopters—who are willing to innovate before the rest of their peers—and there are so-called laggards, who for various reasons will be the last to adopt (Leonard-Barton & Deschamps, 1988). Infrastructure supporting the innovation is very important within corporations (Culnan, 1984; Johnson & Rice, 1987), as it is in general diffusion situations (Brown, 1981; Walton, 1987). Diffusion within organizations also differs from the general model in some important ways. People do not have the same freedom of choice in their roles as corporate employees as they do outside those organizations. Influences within the corporation include whether or not use is mandated by management and the extent to which the innovation is supported by the organizational reward systems.

User Involvement in Design/Development

User involvement in technology transfer ranges from superficial input late in the development process, when only minor alterations to the technology could be considered, to actual development partnership between technology sources and receivers in the construction of innovations from the beginning.

Motives for User Involvement

In either transfer mode, two motives inspire technology developers to involve the intended users or receivers of a new technology in its development: (1) acquiring knowledge from the user/receivers needed in the development process so as to create value, that is, a relative advantage over previous practices, and to ensure usability (Boland, 1978); and (2) attaining user "buy-in"—that is, user acceptance of the innovation and commitment to use it because, as one developer noted, "involvement improves acceptance" (Ives & Olsen, 1984).

The first motive rests on the assumption that the interaction between the technology source and receivers during development is synergistic; that is, this interaction creates a better design than either group could have generated alone (Whitney, 1988). Such a premise similarly underlies the use of "lead users" of current technologies by market researchers to inform the design of the next generation of that product (von Hippel, 1988).

The second reason for user involvement (attaining buy-in) is independent of the first, because increasing user acceptance does not always improve the innovation's quality (Doktor, Schultz, & Slevin, 1979; Ives & Olsen, 1984). When developers involve users to achieve acceptance, they expect them to generate favorable information about the innovation and to be more tolerant of initial inadequacies because of their sense of ownership. When users are involved heavily in development, they tend to take the perspective expressed by a user manager who was interviewed in the present research: "It's very difficult to say we can't accept stuff that we developed jointly."

Point-to-Point: Direct User Involvement

In a point-to-point transfer, it is at least conceivable that all of the actual potential receivers of the technology can be involved—in one seamless process from design through implementation. In fact, in a number of the cases studied, technology sources and technology receivers worked as teams; design and implementation were not separate activities. The higher the proportion of user/receivers involved, however, and the more substantial their contributions to the transfer and (in particular) the design effort, the greater must be the willingness of the receiving management to take such an active role. Users do not always want to be involved in development; a number of developers in the present research lamented that their users were uninterested in any involvement prior to a finished product. Very few users took the position that the key to success was early involvement.

If direct involvement of user/receivers was difficult because of geography or lack of interest, developers in point-to-point transfers occasionally employed surrogate users, that is, persons who once held the same positions as the user/receivers and therefore had process knowledge and could represent the user/receiver viewpoint. A surrogate user is valuable during design, but unless that individual returns to the user organization as a missionary for the new technology, this form of user involvement does not create user buy-in. In order to obtain both knowledge and buy-in, technology developers often have to persuade the receiving organization that it is in their best interests to invest the time and effort of collaborative work. As a developer in a project that failed to get early user input noted, "The project confirmed my experience that we need to be more involved with the using customer."

Diffusion: User Involvement

In the diffusion situation, it clearly is impossible to involve all potential technology receivers in the development process. Although the management of the receiving organization still must be persuaded to invest some effort in the technology design phases, the request may not seem so large, because the proportion of the total targeted population of receivers who must devote time and other resources to the effort is smaller. The more pressing managerial issue in the diffusion situation is how to select representatives to involve in the development so that all important interests and needs will be represented.

In diffusion, the development activity is usually separated from implementation, and the two stages are necessarily much more distinct than need be true in point-to-point transfers. Representativeness is particularly important during knowledge acquisition so that the resultant design will be functional. If the main objective is to secure user/receiver acceptance, however, the user representatives involved in implementation must be credible spokespersons in their own organizations so that when they promote the innovation, they are believed and emulated as opinion leaders (Rogers, 1983).

Diffusion: Direct User Involvement

As in the case of point-to-point transfers, representation may be direct (actual users) or indirect (representatives or surrogates). If the objective is acquiring knowledge about user operations to inform the design process, then the developers need to select people possessing the appropriate kind of knowledge. In several cases studied, "user design groups" selected from the ranks of actual end-users were chosen to represent the different needs of the potential users. Management of such groups is a difficult task (Mumford & MacDonald, 1989) because although receivers with deep process knowledge are good informants on basic design issues, their atypically high level of technical knowledge may render them poor judges of receiver/user interface needs. Observers of a very difficult transfer commented that the developers involved "the wrong end-user/customer; they got answers from senior management instead of from the guy in the trenches."

Moreover, users representing different needs (such as those of experts at the task being enhanced versus those of novices) may not work well together (Leonard-Barton, 1987). Therefore, it is often necessary for developers to convene more than one user design group in order to represent the targeted population, both horizontally (i.e., across user/receiver groups from different organizational departments) and vertically (in terms of technical skill levels). Finally, when user representatives work closely with developers over a long period of time, they may so absorb the developer viewpoint that they cease to represent that of the users (Leonard-Barton, 1987). This problem is akin to that of the "inauthentic professionalism" noted by Rogers (1983, p. 331) when change-agent aides destroy "the very heterophily-bridging function for which [they] were employed." User design groups, however, do help developers make important design trade-offs; therefore, managers in the cases studied generally endorsed the practice.

Recent advances in computer networking technologies have made possible the formation of virtual user/receiver design groups that are never convened physically and therefore have no group identity, but whose members interact as individuals with technology developers to have significant impact on the design of the technology. For example, a computer conferencing system was launched in a computer company with the expectation—fully realized—that users would take advantage of a file established to receive user suggestions. Such virtual user groups do not emulate perfectly the actual group meetings that can be held between developers and users, in that feedback is lagged somewhat and group members usually do not interact directly with each other. Their comments, however, are available to anyone involved with the computer "conference" about the new technology, so that the information is shared to some extent.

Besides the obvious advantage of costing less money than face-to-face meetings among geographically dispersed individuals, such virtual groups have several other benefits. They can represent different levels of technical skill without the dampening influence of small-group dynamics, through which a few powerful or particularly articulate individuals can dominate. They also can draw upon a larger number of potential users than face-to-face meetings. Disadvantages are that developers have little overt power over such groups, and dedication is lessened by the physical and psychological distance from the developers. Such groups have worked very well, however, in organizations in which the needed technology skills are widespread.

Diffusion: Indirect User Involvement

A more indirect way of representing users in a diffusion process is to designate a small buffer group or an individual to obtain design suggestions and desires systematically from all known potential user sites and, using this input, to design a very generic tool to be used by all. This description may sound like the job of a traditional systems analyst. Like the surrogate users described above, however, such individuals often have expertise in the operations of the user/receivers beyond that which might be expected of analysts. This knowledge enables them to be effective "integrators" (Daft & Lengel, 1986). Moreover, unlike the surrogate users, such integrators often have implementation as well as development responsibilities.

In the cases studied,[1] the development and implementation strategies that were followed in diffusion situations were influenced by the corporate cultures of the five firms, see Table 2.2. The more hierarchically organized and managed companies tended to designate buffer development groups (such as those just described) to serve as clearinghouses for all design suggestions about each major tool project. In the much more decentralized and less structured corporations, developers were more likely to work in a two-step process: first identifying and working with one group of potential users who were willing to invest the effort in the software development process, and then diffusing the package to other user units.

Both strategies have potential disadvantages. When buffer groups were used, not all user/receivers felt that the final generic design was appropriate for their particular needs. "They talked generic," said one group about the corporate developers, "but they focused on the equipment used in the major plants funding them." The disadvantage of the second strategy was that each tool's usefulness was extremely dependent on selection of the initial site.

TABLE 2.2: Frequency of Research Cases in Study C by Each Transfer Mode

Transfer Mode	Number of Cases	Exemplary Case
Simple Point-to-Point	11	ROAD, an expert system (form of artificial intelligence that mimics human judgment) functioning as a decision-support tool for a single dispatcher who schedules shipments of computer components by truck around the nation. The system helps to optimize the carrier contracts for efficiency and to reduce cost.
Complex Point-to-Point	3	SCHEDULER, a real-time expert system that monitors work in process on the factory floor and matches demand for materials with capacity on the line. It is being used at one location by about 20 operators, manufacturing quality personnel, and process engineering.
Simple Diffusion	15	SIM, a tool for computer hardware designers that enables representation, simulation, and verification of a hardware design. It is used in about 200 sites by more than 1,300 circuit design engineers.
Complex Diffusion	5	A parts information database used by more than 4,000 engineers and production control, purchasing, customs, and component qualification personnel at 120 sites worldwide.
Total Number of Cases	34	

This point was illustrated in a comparative study of three plants implementing a software package designed by corporate services to automate and monitor purchasing functions within manufacturing (see Note 1). A "beta" site plant selected as the first recipient of the package differed from the others in several important ways, the most important of which was its atypically low number of purchased parts with long lead times. Based on its experience with this nonrepresentative plant, the corporate team programmed the software to order this category of parts to arrive only once in six months.

When the software was implemented in other plants' purchasing departments, where as many as 40% of the orders fell into the long lead-time category, the receiving departments quite literally were buried with incoming components on a given day every six months. This apparently simple miscalculation—with its attendant complications—was very difficult to correct locally at the individual plants, for they did not own or have access to the centrally controlled software source code that needed to be reprogrammed. The needed adjustments took well over one year. As this example suggests, the more that a new technology is customized for the first site using it, the more important are the criteria used to select that site.

Similar problems occur when a technology initially transferred point-to-point, with all the usual customization to the needs of a specific group of users, is diffused afterward to other sites. Such transfers work only if both task and user/receiver skills in the secondary site match those in the first. For instance, 1 of the 17 expert systems studied (so-called because the software mimics the performance of human experts, using an artificial-intelligence approach) was devised by expert factory floor technicians, with minimal help from computer scientists. Because this small diagnostic test system, ADEPT, was designed by representatives of the user population and spoke "their language," it was well accepted in the United States. Technicians in a Japanese sister plant building the same electronic product, however, rejected ADEPT as too elementary; they were too experienced to need it.

In general, managers who know they will be diffusing a technology attempt to avoid customization. In fact, the degree of customization tends to vary according to the intended mode of transfer. As Figure 2.3 suggests, transfers located off the diagonal risk unnecessary costs and inefficiencies. That is, developing a technology with widely applicable generic characteristics when it is to be used by only one site could involve unnecessary research into the needs or desires of irrelevant receivers and possible inefficiencies in site operations once the tool is installed. On the other hand, heavily customizing a technology to the needs of a single site when it is to be diffused widely can be very expensive, as each site will require costly redesigns.

Pressures to customize create conflict for tool developers within organizations because they are internal vendors for multiple sites. Internal customers feel their fellow employees should be even more responsive to their needs than external vendors would be: "After all, they (the developers) are in the business of making my life simpler," one user observed.

All such problems encountered in the two transfer modes are exacerbated in the complex forms of those modes. For instance, although developers are always under pressure to customize in point-to-point

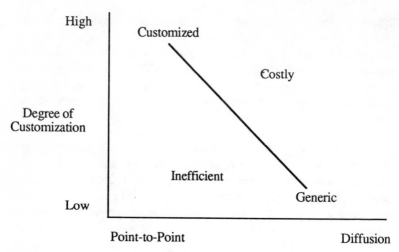

Figure 2.3. Customization and Modes of Transfer

transfers, the problem is worse when the tool is to be used by people performing different tasks—often in highly differentiated departments, with conflicting priorities and rewards (Lawrence & Lorsch, 1967). Consider, for instance, a complex point-to-point transfer of a manufacturing resource-planning system (MRPII) to a single factory. The software coordinates the operations of multiple departments within the factory, and each department (e.g., purchasing, materials handling, and engineering) accesses a different set of modules in the software, albeit with some overlapping. Hardware can introduce similar complexity, as in the case of a piece of test equipment that can be used in manufacturing to check the progress of hardware during production—and also in the field by service technicians to diagnose problems. Such tools are far more complex to transfer, as well as to design, than a tool that serves a single job category.

The level of complexity is ratcheted up yet another notch in the case of complex diffusion, when multiple groups of diverse users must be served. An example is the diffusion of an MRPII system such as the one mentioned above to all the factories in a large corporation. One of the engineering design tool cases studied, Millium, was similarly complex. Millium allows both hardware and software engineers to design and simulate the user interface on a piece of office equipment (i.e., the buttons and displays used to operate the machine). Each of the multiple groups of industrial engineers and product designers working on various models of the product line needed slightly different functions and interfaces with the tool. In this case, the groups constructed their own, resulting in dozens of versions of Millium. When the time came to update and enhance it to handle new design problems, no one group owned the responsibility.

Managing the Different Transfer Situations

Each of the four transfer situations encountered calls for somewhat different managerial skills and attention to somewhat different critical success factors. In the simple point-to-point transfer, face-to-face negotiation about the technology design and use is common throughout the process. Because the user/receivers can be involved directly—and heavily—in the design and can make design trade-offs directly, the resulting technology often is very customized.

One of the primary dangers for developers to avoid is automating history, that is, fitting the technology to exactly what the user/receivers demand today, with little thought for the future. Often users lack the technical understanding to have such vision: "They don't ask for the system because they don't know it is possible." Faced with these constraints, some developers take the strategy of giving users "more than they wanted," so that they can grow into the technology.

In complex point-to-point transfers, the users again can be involved directly, but there are multiple stakeholders among the user/receiver groups, with separate and sometimes conflicting needs. Here again, the manager calls upon negotiation skills to avoid optimizing for one special group at the expense of the whole, often interlinked systems of user/receiver organizations. One solution to this dilemma is the deliberate separation and protection of a core functionality in the software tool. The core is surrounded with very generic modules that can be customized by each separate user/receiver group.

Diffusion requires a much more structured approach, akin to marketing a new product. In the simple diffusion cases, the development groups conceived of archetypical customers (e.g., a design engineer or a test technician) and designed for that specialized market. The more successful groups, however, also recognized the need to accommodate variance in skill levels among this targeted group. They built alternative user interfaces, so that users with different skill levels could interact with the software at their own level. This design strategy is illustrated by an expert system (SNAIL) that monitors, diagnoses problems, and suggests solutions in continuous-flow manufacturing processes (such as oil refineries or paper mills). Three different user interfaces are available: novice, regular, or expert. Users can (1) receive advice about actions to follow along with an explanation of the reasoning leading to the suggestion (novice); (2) receive advice routinely, but without the explanation (regular); or (3) follow actions without any advice, unless the system recognizes a potential error or the operator fails to follow expected actions, in which case the software activates itself (expert).

In the case of complex diffusion, the developers must work with multiple archetypes of user/receivers and therefore aim at the most generic market possible. For instance, developers creating the new conferencing system (mentioned above) at a large computer company used an already existing and widely used electronic mail system as a familiar interface. Both this interface and the conferencing system into which it tapped were extremely generic in nature, customized for no one group.

The primary managerial skills required in point-to-point transfers are negotiating skills: understanding how to overcome the initial positions (i.e., confrontational stances about means to an end) of various groups by translating them into interests (i.e., explanations of desired ends) so that common ground may be found (Fisher & Ury, 1981). Implementation during point-to-point transfers occasions much mutual adaptation of both the technology and the receiving organization (Leonard-Barton, 1988a). The more numerous the receiving groups—that is, the more complex the point-to-point transfer—the more interpersonal and negotiating skills the project manager needs.

The manager of a diffusion situation, who is more akin to a marketer than a negotiator, requires skills in organizing as much as interpersonal ones. Although the effort required to prepare a tool for transfer (e.g., writing formal documentation, organizing formal training sessions) might be desirable even for one user site, nevertheless, the fewer the users, the less likely it is that developers would be willing to spend the needed time and effort.

The diffusion manager's tasks are similar to those involved in delivering a new product to market, with concern for all the steps, from advertising to distribution and field support of the innovation (Gatignon & Robertson, 1985; Leonard-Barton & Kraus, 1985; Rogers, 1983). Awareness of an innovation can be raised through formal advertising and written media, even within corporations (e.g., through in-house newsletters). Persuasive articles (e.g., written with early users or about a demonstration project) can be used to sell the idea to the target user population, or widespread memos and directives can be sent out to mandate use.

This attention to the implementation phase in diffusion modes of technology transfer require enormously careful coordination, and it consumes a lot of resources. These are some of the reasons why the manager charged with development or implementation of a new technology in a corporation would wish to be aware from the beginning of whether the transfer is a point-to-point or diffusion situation.

Notes

1. The 51 cases upon which this chapter draws for illustrative material were grouped in three separate but complementary studies of the same phenomenon, namely, the development and internal transfer of technologies used to produce goods or services from the originating laboratory or development group to operations. These studies were conducted during the years 1983 to 1989.

   A. 10 cases, including a three-year longitudinal real-time study of XSEL (Leonard-Barton, 1987) and nine retrospective cases of various hardware, software, and materials innovations (Leonard-Barton, 1988a,b). See Leonard-Barton (1990a) for a description of the methodology employed.

   B. A comparative study of two very similar software packages used in the purchasing departments of manufacturing plants in two competitor corporations (including one of the same corporations as in study A). In each corporation, relatively successful and unsuccessful implementations of the package were studied in three plant sites, for a total of six cases (Leonard-Barton, 1989).

   C. In four *Fortune 100* electronics firms, a total of 34 software development projects (Table 2.2). Each software package was developed internally for internal use in engineering or manufacturing. In each case, at least the following informants were interviewed: manager of the development project, a key developer, manager of the using work unit, and a supervisor or senior operator in the using work unit. A total of more than 155 interviews were conducted, using semistructured interview protocols.

2. There has been much debate about the connection between technology and organizational structure (Fry, 1982). Gerwin (1981) distinguishes between technology and structure but maintains that "there is no clear-cut distinction between tasks and technology; rather, there is a more or less gradual shift from ends to means" (p. 5). He therefore

discusses the "task technology combination" and points out that technology may determine task, as well as the reverse.

3. The concepts presented here are applicable also to such situations as the transfer of technology from one research laboratory to another; for instance, from a laboratory at a university to one at a corporation. Such transfers, however, are relatively less important in this discussion because they are characterized predominantly by a single mode: point-to-point. Transfer from a jointly funded research consortium such as MCC to the laboratories of its corporate sponsors falls in the more difficult, complex point-to-point mode.

4. Characterization of the mode thus depends in part on the boundaries used to define the managerial situation. If the development group makes multiple, sequential transfers to many different groups, the whole process over time could be considered a diffusion problem. If each transaction is considered separately, however, it is a point-to-point transfer. If the development group handles a limited number of such transfers simultaneously, the process falls somewhere between the two extremes.

5. For earlier formulations of these concepts of span and scope leading to implementation complexity, see Leonard-Barton (1988b).

6. The interesting cases of "reinvention," when users adapt tools to their own unique purposes—which are often quite different from the original intent and which therefore result in one tool being applied to multiple tasks (Rogers, Eveland, & Klepper, 1977)—are not included here.

References

Boland, R. J. (1978). The processes and product of system design. *Management Science, 24*(9), 887-898.

Bradbury, F., Jervis, P., Johnston, R., & Pearson, A. (1978). *Transfer process in technical change.* Alphen aan den Rijn, Netherlands: Sijthoff & Noordhoff.

Brown, L. A. (1981). *Innovation diffusion: A new perspective.* New York: Methuen.

Culnan, M. J. (1984, April). The dimensions of accessibility to online information: Implications for implementing office information systems. *ACM Transactions on Office Information Systems, 2*(2), 141-150.

Czepiel, J. A. (1974). Patterns of interorganizational communications and the diffusion of a major technological innovation in a competitive industrial community. *Academy of Management Journal, 18,* 6-24.

Daft, R. L., & Lengel, R. H. (1986). Organizational information requirements, media richness and structural design. *Management Science, 32*(5), 554-571.

Doktor, R., Schultz, R. L., & Slevin, D. P. (1979). *The implementations of management science* (Studies in Management Sciences, Vol. 13). New York: North-Holland.

Fisher, R., & Ury, W. (1981). *Getting to yes: Negotiating agreement without giving in.* Boston: Houghton Mifflin.

Fry, L. W. (1982). Technology-structure research: Three critical issues. *Academy of Management Journal, 25*(3), 532-552.

Gatignon, H., & Robertson, T. S. (1985, March). A propositional inventory of new diffusion research. *Journal of Consumer Research, 11,* 849-867.

Gerwin, D. (1981). Relationships between structure and technology. In P. C. Nystrom & W. Starbuck (Eds.), *Handbook of organizational design, vol. 2.* Oxford: Oxford University Press.

Ives, B., & Olsen, M. H. (1984, May). User involvement and MIS success: A review of research. *Management Science 30*(5), 586-603.

Johnson, B., & Rice, R. (1987). *Managing organizational innovation: The evolution from word processing to office information systems.* New York: Columbia University Press.

Lawrence, P. R., & Lorsch, J. W. (1967), *Organization and environment.* Boston: Harvard Business School Press.

Leonard-Barton, D. (1987, fall). *The case for integrative innovation: An expert system at Digital.* Sloan Management Review, 29(1), 7-19.

Leonard-Barton, D. (1988a). Implementing as mutual adaptation of technology and organization. *Research Policy, 17,* 251-267.

Leonard-Barton, D. (1988b). Implementation characteristics of organizational innovations. *Communication Research, Vol. 15*(5), 603-631.

Leonard-Barton, D. (1989). Implementing new production technologies: Exercises in corporate learning. In M. A. von Glinow & S. Mohrman (Eds.), *Managing complexity in high technology industries: Systems and people.* New York: Oxford University Press.

Leonard-Barton, D. (1990). A dual methodology for case studies: Synergistic use of a single site with replicated multiple sites. *Organization Science, 1*(3).

Leonard-Barton, D., & Kraus, W. A. (1985). "Implementing new technology." *Harvard Business Review,* November-December, pp. 102-110. Reprint #85612.

Leonard-Barton, D., & Rogers, E. (1981). *Horizontal diffusion of innovations: An alternative paradigm to the classical model.* Working Paper 1214-81, Sloan School of Management, MIT, Cambridge, MA.

Leonard-Barton, D., & Deschamps, I. (1988, October). Managerial influences in the implementation of a new technology. *Management Science, 34*(10), 1252-1265.

Mansfield, E., et al. (1977). *The production and application of new industrial technology.* New York: Norton.

Mansfield, E., et al. (1982). *Technology transfer, productivity and economic policy.* New York: Norton.

Midgley, D. (1977). *Innovation and new product marketing.* New York: Wiley.

Mumford, E., & MacDonald, W. B. *Xsel's progress: The continuing journey of an expert system.* New York: Wiley.

Perrow, C. (1967). A framework for the comparative analysis of organizations. *American Sociological Review, 32,* 194-208.

Rogers, E. M. (1983). *Diffusion of innovations* (3rd ed.). New York: Free Press.

Rogers, E. M., Eveland, J. D., & Klepper, C. A. (1977). *The innovation process in public organizations: Some elements of a preliminary model.* Final report to National Science Foundation (Grant RDA 75-177952).

Rousseau, D. M., & Cooke, R. A. (1984). Technology and structure: The concrete, abstract, and activity systems of organizations. *Journal of Management, 10*(3), 345-361.

Rubenstein, A. H. (1989). *Managing technology in the decentralized firm.* New York: Wiley.

Schon, D. A. (1967). *Technology and change: The new Heraclitus.* New York: Delacorte.

Souder, W. (1987). *Managing new product innovations.* Lexington, MA: Lexington.

Ulrich, R. A., & Weiland, G. F. (1980). *Organization design and theory.* Homewood, IL: Irwin.

von Hippel, E. (1988). *The sources of innovation*. New York: Oxford University Press.

Walton, R. E. (1987). *Innovating to compete*. San Francisco: Jossey-Bass.

Whitney, D. E. (1988). Manufacturing by design. *Harvard Business Review, 66*(4), 83-91.

Woodward, J. (1965). *Industrial organization: Theory and practice*. London: Oxford University Press.

3

The Interorganizational Environment: Network Theory, Tools, and Applications

ELLEN R. AUSTER

International interorganizational relationships have exploded in the last decade ("Business Without Borders," 1988) and, not surprisingly, researchers increasingly have become interested in this phenomenon. A strong foundation of research grounded in strategic and transaction cost approaches is emerging on dyads at the organizational and industry level. This chapter proposes that a network perspective—which considers the constellation of relationships these dyads are embedded in—could enhance this research stream by bringing issues of power, resource dependence,

AUTHOR'S NOTE: The cooperation of top managers interviewed in the United States and Japan and the Japanese External Trade Organization is acknowledged gratefully. My thanks to Howard Aldrich, Don Beard, Warren Boeker, David Gibson, Don Hambrick, Bill McKelvey, Mike Tushman, Paul Olk, and Steve Weiss for their comments and suggestions on earlier drafts. Support for this research was provided by the Center for Japanese Economy and Business at Columbia University and a Columbia University Graduate School of Business Research Fellowship.

*and exchange to the forefront. The author concentrates on and
contributes to the macro literature on international interorganizational relationships. Data on U.S.-Japan linkages are
used to illustrate a network approach, and implications for future
research on interorganizational relationships at multiple levels of
analysis are considered. Ellen R. Auster is an Associate Professor
in the Graduate School of Business at Columbia University.*

An Interorganizational Network Approach

In rapidly changing technological and market environments, international interorganizational relationships offer a means to diversify
cost and risk and co-opt or block competition while gaining access to
new technologies, customers, products, distribution channels, and resources (Auster, 1987; Berlew, 1984; Contractor & Lorange, 1988;
Harrigan, 1987; Killing, 1983). Early macro research on international
interorganizational relationships focused primarily on one type of interfirm linkage, joint ventures, and was based mostly in strategy
(Harrigan, 1985b; Killing, 1983). The competitive benefits of alliances
in improving the firm's strategic posture in its industry and issues in
managing the "parent/child/parent" relationships were the topics of
greatest concern (Bivens & Lovell, 1966; Harrigan, 1985).

Recent macro research is more diverse. The structural and process
issues examined in interorganizational relationships have broadened to
include topics ranging from what types of human resource programs are
most effective in international joint ventures to the relationship of
industry conditions to joint-venture performance (Hladik, 1988; Pucik,
1988). Rather than an exclusive focus on joint ventures, multiple forms
of linkages such as technological transfers and joint R&D are analyzed
and compared to joint ventures (Doz, 1988; Hergert & Morris, 1987;
Pisano, Russo, & Teece, 1988; Pisano, Shan, & Teece, 1988). Transaction cost approaches often are used as an alternative or complementary
theoretical framework (Brahm & Astley, 1988; Kogut & Singh, 1988;
Pisano, Russo, & Teece, 1988; Pisano, Shan, & Teece, 1988). Arguing
that interorganizational relationships are contractual agreements that
stand in between the typical make-or-buy, hierarchy-or-market distinc-

tions, these studies have focused on issues at both the organization and industry levels of analysis.

Organizational-level studies have explored issues such as choices of governance modes in joint ventures and their impact on performance (Brahm & Astley, 1988). Industry level studies have examined formation rates of different types of interorganizational linkages within biotechnology and telecommunications (Pisano, Russo, & Teece, 1988; Pisano, Shan, & Teece, 1988) or choice of entry mode as it relates to variables such as industry R&D intensity, marketing intensity, growth, and concentration (Kogut & Singh, 1988).

The conceptual and analytical tools of network methodology combined with exchange and resource dependence theoretical perspectives offer a powerful, useful, and insight-provoking framework for enriching this research stream. It directs our attention to important but relatively neglected dimensions of these relationships such as power, reciprocity, influence, and interdependency. Strategic and transaction cost approaches tend to emphasize economic costs and benefits—often in the shorter run—and typically analyze dyads or triads (parent/child/parent) of relationships. Power dynamics can be inferred or extracted from some variables, such as asset specificity or sunk costs, but often power dynamics or the implications of webs of relationships are overlooked.

Network analysis, in contrast, brings to the surface the webs of relationships that these dyads and triads are embedded in. The imagery changes from a focus on pairs of partners and isolated linkages to one where constellations, wheels, and systems of relationships are examined (Rogers & Kincaid, 1981). In doing so, new angles, questions, and insights emerge.

There are pragmatic as well as analytical reasons for expanding our frameworks to include networks. Consider these excerpts from interviews recently conducted with top managers in Japan at a number of companies including Hitachi, Toshiba, and Mitsubishi Electric (personal interviews, March, 1989).

- "When I evaluate a possible joint venture or technological cooperation, I talk to other Japanese companies and ask what their experiences have been with that U.S. company."
- "I am aware of most of the relationships our company and its sister companies have with U.S. companies. This enables me to see the larger picture."

- "Any particular venture can only be evaluated by considering the other ventures we have with that company and with others in the industry."

What is striking about these comments is the consideration of portfolios of interorganizational relationships and how they are connected to the interdependence within the industry. It is clear that an evaluation of any specific interorganizational linkage involves an assessment of how this connection meshes with other linkages already established in the industry. And networks are used to acquire information, for both now and in the future. The importance of a network orientation to the Japanese is indicated further by the resources devoted to tracking interdependencies. One branch of the Ministry of Trade and Industry (the Japanese External Trade Organization) regularly tracks all forms of linkages announced publicly. In addition, many companies have their own in-house tracking and monitoring of interorganizational hookups in their industry.

In the United States, in contrast, some public information is available on joint ventures with foreign companies, but systematically collected information on other forms of international hookups such as joint R&D or technological transfers is not accessed easily and is only beginning to be compiled, often by academics. More surprising perhaps is the lack of in-house information collected by companies tracking the interdependencies within their industries. Many companies interviewed did not even have accurate reports of current interorganizational linkages with Japanese companies, because the decisions and information often remain at the business-unit level and are not centralized.

The field of international interorganizational studies has grown tremendously in the last decade. Multiple forms of relationships are being examined. Research questions, grounded in transaction cost and strategic views, address issues at both the organizational level of analysis and the industry level of analysis. Moreover, studies are exploring both the process side of managing these linkages and structural features, such as the distribution of forms within industries or industry conditions and their relationship to formation, mortality, and performance. The research orientation, however, like that of many U.S. managers, has tended to be dyadic with an emphasis on costs and benefits in the short run.[1] A network perspective beckons consideration of issues such as power and dependence in the long run by bringing the webs of relationships in which firms are embedded into focus.

Theoretical Underpinnings of a Network Approach

A network can be defined as all the linkages between actors in a system (Rogers & Kincaid, 1981). In its essence, network analysis is an analytical tool, not a theoretical framework. Since the mid-1970s, however, network analysis has been grounded primarily in theories of exchange, power, and resource dependence (Aldrich, 1979; Aldrich & Whetten, 1981; Cook, 1982; Pfeffer & Salancik, 1978). Although initially somewhat distinctive approaches, these three perspectives have merged in recent years as their overlaps have become more apparent. Power can be viewed as asymmetric exchanges, and dependence the outcome of exerted power. The key assumptions underlying this network approach can be summarized as follows:

- Actors attempt to establish linkages in order to acquire resources or information about their environment, coordinate competitive interdependence, or reduce competitive uncertainty and thereby increase their power (Pfeffer & Salancik, 1978).
- Action is viewed as intentional; thus, ties between actors are established, maintained, or broken because of their perceived value (Cook, 1982).
- Networks represent interconnected flows of resources and resource dependencies (power relationships) between actors. This flow and its causes and consequences are the focus of network analysis (Aldrich, 1979; Cook, 1982; Pfeffer & Salancik, 1978).
- Networks are dynamic; their configurations shift and change as actors attempt to gain or balance power by redistributing resources (Cook, 1982).

Thus, from a network perspective, linkages are formed intentionally in order to manage uncertainty and acquire resources, information, and power. Networks are systems of these resource dependencies that are dynamic as a result of the actions of the actors involved.

The Analytical Tools of Network Analysis

Different research questions require different tools. This section provides an overview and brief discussion of a variety of tools that may be useful for analyzing a range of questions pertinent to international interorganizational relationships.

Many different structural dimensions of networks have been developed by sociologists and social psychologists studying networks of individuals (Burt, Minor, & Associates, 1983; Rogers & Kincaid, 1981). Some have been applied to the study of domestic interorganizational relations (see, for example, Aldrich & Whetten, 1981; Cook, 1977). The literature on networks, however, is not easily accessible. As Burt (1980) notes, "Anyone reading through what purports to be a 'network' literature will readily perceive . . . the analogy between that literature" (p. 79) and what Barnes (1972) labeled "a terminological jungle in which any newcomer may plant a tree." The purpose of this discussion is to thrash through that terminological jungle, simplifying it where possible and extracting components that are useful for understanding networks of organizations, rather than networks of people.

Network Boundaries

Establishing the boundaries of the networks is the first issue confronted when conducting network analyses. How the boundaries of networks are drawn is a critical step, for it creates the sample of linkages that are examined further. The boundaries of a system may be defined by the researcher (*nominalist approach*) or constructed socially by those involved (*realist approach*) (Laumann, Marsden, & Prensky, 1983). In the nominalist approach, the selection criteria for organizations chosen for a network might be based on attributes of the organizations, activities the organizations are involved in, or characteristics of their relations. Defining a network by its home country, size or age of the organization, or whether it is a Fortune 500 company would be examples of selection based on attributes.

Choosing companies based on the types of products or services they produce would be an activity-based method of selection. Characteristics of relations as a method of selection might mean a focus on companies currently engaged in overseas joint ventures. Often many different selection criteria will be combined to define the boundaries of the system. The realist approach uses the social construction of those involved to define the boundaries of the system. This approach, however, does have nominalist qualities. Those included in the sample reflect the researcher's selection criteria and expected boundaries of the system.

The Building Blocks of Networks:
Interorganizational Linkages

Interorganizational linkages are the building blocks of networks and can be defined as relations between two or more organizations formed to transfer, exchange, develop, or produce technology, raw materials, products, or information. The term *linkage* rather than alliance, collaborative agreement, or cooperation is used because the extent to which these relationships are mutually beneficial should not be overestimated. The firms may in fact be exploiting each other and pursuing contradictory goals, and their purposes may change over time (Auster, 1987; Buckley & Casson, 1988).

Understanding the basic underlying structure connecting linkages is important in any network analysis. Relationships in a linkage may be one-way (also called *asymmetric,* or *unilateral*) or two-way (also called *reciprocal, symmetric,* or *bilateral*), horizontal or vertical. Horizontal linkages refer to exchanges between organizations producing similar products, processes, or resources. This type of relation has also been called *commensalistic* (Aldrich, 1979, p. 266; Hawley, 1950, p. 39) or described as interdependence in the same stage of the value or transformation chain (Harrigan, 1985b; Porter, 1985). Vertical linkages refer to exchanges between organizations at different stages of the production and distribution chain (Contractor & Lorange, 1988; Pennings, 1981; Porter, 1985). GM's joint venture with Akebono Brake would be an example; these organizations have complementary relations with each other "in production or the rendering of services to clients" (Pennings, 1981, p. 434). Backward vertical interdependence refers to a linkage with a firm in an earlier stage of the transformation process. Forward vertical interdependence refers to linking up with a firm at a later stage of the transformation process.

The content of a linkage refers to what is exchanged or transmitted (Blau, 1964; Homans, 1961). In the sociological and psychological network literature, studies have analyzed content relations such as friendship, acquaintance, work, kinship, and intimacy (Burt et al., 1983; Rogers & Kincaid, 1981). In the interorganizational context, some common linkages include: OEM supply linkages, licensing, technological transfers or exchanges, joint research and development, and joint ventures (Table 3.1).

The degree of dependence varies depending on the content of linkages, as several researchers have noted recently. Contractor and

TABLE 3.1: The Generic Content and Structure of Common Interorganizational Linkages

OEM Supply	— a one-way linkage formed to sell raw materials or products from organization X to organization Y
Licensing	— organization Y buys the right to use a process or product for a limited time period from organization X
Technological transfer	— a one-way linkage formed to transfer technology from organization X to organization Y
Technological exchange	— a two-way linkage formed to exchange technology or technological information between organization X and organization Y
Joint R&D	— a two-way linkage formed to develop and share research jointly between organization X and organization Y
Joint venture	— organization X and organization Y create a separate organizational entity to produce goods or services

Lorange (1988), for example, develop a continuum of interorganizational dependence in linkages based on the type of compensation between partners. Technological training is the lowest on their scale because compensation is based on a lump-sum fee. Joint ventures anchor the high end of the scale with compensation based on a fraction of shares or dividends. In between these extremes are forms such as production agreements with compensation based on a markup on components sold or finished, and licensing with compensation based on royalties.

Auster (1990) offers an alternative scheme by using the degree of resource investment to rank the relative dependence of forms. Thus, LRILs (low resource investment linkages) would include relationships such as technological transfers and joint R&D that are more autonomous and severed more easily. High resource investment linkages (HRILs) would be forms such as joint ventures that require much longer-term commitment and trust, a great financial investment, the construction or acquisition of a space to house the venture, equipment and technology to produce the output, and more management time and energy to oversee the venture. Given this high resource investment, switching costs and barriers to exit are formidable obstacles to termination.

Having established the boundaries of the system and delineated common structures and contents of the linkages within the system, it is appropriate to turn to tools for analyzing networks. Table 3.2 operationally defines the terms discussed in subsequent sections. Figure 3.1

TABLE 3.2: Structural Dimensions of Networks

Network as Focus of Analysis

Size	— number of organizations in the network*
Density	— number of linkages in the network*
Diversity	— linkage: number of different types of linkages in the network*
	— organizational: the number of different types of organizations in the network*
Reachability	— the number of links separating two organizations
Stability	— linkage: whether the form of linkage in the network remains the same over time*
	— organizational: whether the organizations in the network remain the same over time*
frequency of change	— how often linkages or organizations change*
magnitude of change	— how many linkages or organizations change*
Stars	— the number of organizations with greater than X number of ties
Isolates	— the number of organizations with no linkages to other organizations
Linking pins	— organizations with extensive and overlapping ties to different parts of a network

Organizational Position within a Network as Focus of Analysis

Centrality	— the proportion of the sum of relations that involve organization X
Range	— the number of contacts organization X has
Multiplexity	— the extent that organization X is connected to a high proportion of organizations in the network by multiple types of relations
Degree of Horizontal Interdependence	— the number of linkages with organization X at the same stage of the transformation process
Degree of Vertical Interdependence	— the number of linkages with organization X at different stages of the transformation process
backward	— the number of linkages with organization X at an earlier stage of the transformation process
forward	— the number of linkages with organization X at a later stage in the transformation process.

*This dimension is applicable to organization set analysis.

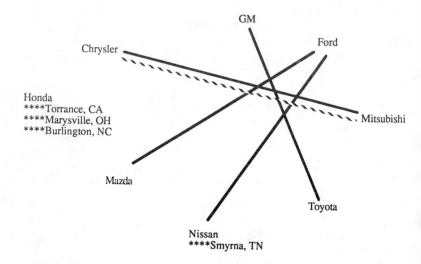

Figure 3.1. Interorganizational Relationships Created Between Top U.S. Auto Companies and Top Japanese Auto Companies

NOTE: All linkages were created in 1984 and 1985.
SOURCE: Based on case data from *Cooperations Between American and Japanese Firms: Cases of Industrial Cooperation in 1984* (New York: Japanese External Trade Organization, pp. 1-154); *Cooperations Between American and Japanese Firms: Cases of Industrial Cooperation in 1985* (New York: Japanese External Trade Organization, pp. 1-278); and *JETRO Monitor*, 3(9), pp. 1-4 (1989).

and Table 3.3 show linkages between the major auto companies in the United States and Japan. Figure 3.2 and Table 3.4 show linkages between the major electronics companies in the United States and Japan. The data for these applications are based on cases of interorganizational linkages and direct investments in the electronics and auto industries that were formed in 1984 and 1985. Direct investment also is displayed because it represents a strategic alternative that many Japanese companies opt for instead of linkage.[2]

TABLE 3.3: A Matrix of Figure 3.1 with Applications of Network Methodology and Concepts

	GM	Ford	Chrysler	Toyota	Nissan	Honda	Mitsubishi	Mazda
GM	0	0	0	1	0	0	0	0
Ford	0	0	0	0	1	0	0	1
Chrysler	0	0	0	0	0	0	2	0
Toyota	1	0	0	0	0	0	0	0
Nissan	0	1	0	0	0	0	0	0
Honda	0	0	0	0	0	0	0	0
Mitsubishi	0	0	1	0	0	0	0	0
Mazda	0	1	0	0	0	0	0	0

Network as Focus
  Size = 8
  Density = 5/8 = 63%
  Diversity = 4
  Reachability = e.g., from General Electric to Westinghouse = 2
  Stars (>2) = 2 (Ford, Chrysler, Mitsubishi)
  Linking Pins = NA
  Isolate = Honda

Organization Position within Network—Ford's position
  Centrality = 40%
  Range = 2
  Multiplexity = Very Low (Ford only has joint ventures)

SOURCE: Based on case data from *Cooperations between American and Japanese Firms: Cases of Industrial Cooperation in 1984* (New York: Japanese External Trade Organization, pp. 1-154) and *Cooperations between American and Japanese Firms: Cases of Industrial Cooperation in 1985* (New York: Japanese External Trade Organization, pp. 1-278) and other data obtained through the Japanese External Trade Organization.

Network approaches fall into basic categories: (1) those analyzing characteristics of networks (networks are the focus of analysis); and (2) those analyzing the position of an organization within a network (an organization's position within the network is the focus of analysis).

Networks as the Focus of Analysis

Networks of organizations can be analyzed in terms of their size, density, diversity, reachability, and stability. These dimensions also can be used as the basis for comparisons across networks. These concepts have been defined and used in many different ways, but the definitions offered below seem most clear and intuitive. The *size* of a network is

TABLE 3.4 A Matrix of Figure 3.2 with Applications of Network Methodology and Concepts

	GE	AT&T	Westinghouse	ITT	Raytheon	Matsushita	Hitachi	Toshiba	NEC	Sony
GE	0	0	0	0	0	1	1	2	1	0
AT&T	0	0	0	0	0	0	0	0	0	0
Westinghouse	0	0	0	0	0	0	0	1	0	0
ITT	0	0	0	0	0	0	0	0	0	0
Raytheon	0	0	0	0	0	0	0	0	0	0
Matsushita	1	0	0	0	0	0	0	0	0	0
Hitachi	1	0	0	0	0	0	0	0	0	0
Toshiba	2	0	1	1	0	0	0	1	0	0
NEC	1	0	0	0	0	0	0	0	0	0
Sony	0	0	0	0	0	0	0	0	0	0

Network as Focus
  Size = 10
  Density = 5/10 = 50%
  Diversity = 4
  Reachability = e.g., General Electric to Westinghouse = 2
  Stars (>3) = 2 (General Electric and Toshiba)
  Isolate = AT&T, ITT, Raytheon, Sony

Organization Position within Network—General Electric's position
  Centrality = 4/10 = 40%
  Range = 4
  Multiplexity = General Electric and Toshiba have higher multiplexity than the other companies

SOURCE: Based on case data from *Cooperations between American and Japanese Firms: Cases of Industrial Cooperation in 1984* (New York: Japanese External Trade Organization, pp. 1-154) and *Cooperations between American and Japanese Firms: Cases of Industrial Cooperation in 1985* (New York: Japanese External Trade Organization, pp. 1-278).

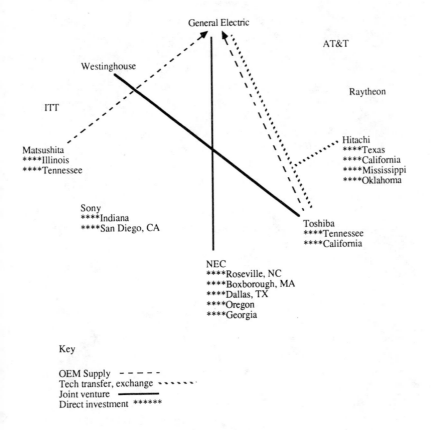

Figure 3.2. Interorganizational Linkages Created between Top U.S. Electronics Companies and Top Japanese Electronics Companies

SOURCE: Based on case data from *Cooperations Between American and Japanese Firms: Cases of Industrial Cooperation in 1984* (New York: Japanese External Trade Organization, pp. 1-154); *Cooperations Between American and Japanese Firms: Cases of Industrial Cooperation in 1985* (New York: Japanese External Trade Organization, pp. 1-278); and *JETRO Monitor*, 3(9), pp. 1-4 (1989).

the number of organizations in the network—in Figure 3.1, size equals 8; in Figure 3.2, size equals 10.

The *density* of a network is the number of linkages in the network (Aldrich & Whetten, 1981). It also can be calculated as a percentage: the number of linkages divided by the size of the network. Or, density can be measured as the number of holes in a matrix where cell *ij* represents the nature of the relationship between organization i and organization j as shown in Table 3.3 and Table 3.4. Figures 3.1 and 3.2 and Tables 3.3 and 3.4 indicate that the auto industry has a higher density than the electronics network (63% versus 50%).

Diversity is the number of different types of linkages found in the network (Burt et al., 1983). In Figures 3.1 and 3.2, there are three major types of linkages shown: OEM supply relationships, technological transfers or exchanges, and joint ventures. *Organizational diversity* is the number of different types of organizations in the network. Diversity could be measured along a number of characteristics including industry or size. *Reachability* is the number of links separating two organizations in the network (Aldrich & Whetten, 1981; Tichy, 1981).

Linkage stability refers to whether the linkages in the network remain the same type over time (Aldrich & Whetten, 1981). For example, the Westinghouse-Toshiba joint venture now is owned wholly by Toshiba. *Organizational stability* could refer to whether the organizations in the network change over time. Two dimensions of stability are: (1) the frequency of change, or how often ties of organizations change; and (2) the magnitude of change, or how many ties change. In the office equipment and computer industry, for example, networks are more fluid than in more mature industries such as autos.

The configuration of networks can also be analyzed both within and across networks. The numbers of stars, isolates, and linking pins can be counted (Rogers & Kincaid, 1981); the number of *stars* would be the number of organizations with greater than x number of ties. If greater than three ties is the defining criterion used in Figure 3.2, then General Electric and Toshiba would be the stars. *Isolates* are those organizations with no linkages to other organizations; Honda Corporation would be an isolate in Figure 3.1 and Table 3.3. Isolates also are created when organizations previously linked in the network become uncoupled over time, such as Toshiba and Westinghouse (as noted above). *Linking pins* are those organizations with extensive and overlapping ties to different parts of a network (Aldrich & Whetten, 1981). General Electric would be a linking pin in Figure 3.2 and Table 3.4.

More sophisticated network techniques, such as structural equivalence, should be noted although space constraints limit the discussion. *Structural equivalence,* strictly speaking, refers to elements in a network that have identical sets of relations (Burt, 1988). It is used typically, however, as a continuous variable based on a calculation of Euclidian distance where 0 equals two perfectly equivalent elements in a network. As the Euclidian distance moves toward 1, the extent to which the elements are involved in different patterns increases. Structural equivalence could be applied at the firm level, for example, in an assessment of whether Mitsui and Mitsubishi engage in similar types of relations with U.S. companies.

Examining the relative position or power of specific organizations is another use of network analysis. Measures of the position of an organization in a network include centrality, range, and multiplexity. *Centrality,* as defined by Burt (1980), is the proportion of the sum of relations within a network that involve actor (organization) x. The centrality ratio for Ford is 40% in Figure 3.1. The absolute number of contacts of actor (organization) x has been called *range* by Burt (1980), although *intensity of contacts* may better convey the meaning of this relationship. Multiplexity for an organization is the extent that organization x is connected to a high proportion of organizations in the network by multiple types of relations (Burt, 1983). General Electric and Toshiba have greater multiplexity than the other companies in their network and higher than any of the companies in the auto network. *Overlap* and *redundancy* (Tichy, 1981; Tichy, Tushman, & Fombrun, 1979) are used as synonyms for multiplexity, but are less useful because they do not capture differences in the content of the linkages.

Applying Network Tools to Organization Sets

A network perspective can be applied to organization sets as well. An *organization set*, as developed by Evan (1963) and Aldrich (1979), is the set of linkages of one focal organization. The linkages of Toshiba, General Electric, and General Motors shown in Figure 3.3 are examples of organization sets of new linkages formed for that time period. Except for notable exceptions such as Evan's (1972) study of federal regulatory commissions or Hirsch's (1972) study of organizational sets in the culture industry, few organizational-set analyses have been done.

Many of the dimensions used for structural analyses of networks can be applied to organization sets. (See Table 3.2 and those terms marked "*".) Size, density, diversity, and stability all tap different features of an organization set that are useful for comparing organization sets of different organizations or of the same organization over time. A comparison of the organization sets in Figure 3.3, for example, reveals several interesting patterns. Toshiba, a high-technology company in

Japan, formed 16 linkages with 14 companies during 1984 and 1985; 25% were joint ventures, 19% were joint R&D, 19% were technological transfers or exchanges, and 38% were OEM supply relationships. General Electric, a large high-technology company in the United States, formed 12 linkages with 10 Japanese companies during the same time period. Of these, 25% were joint ventures, 8% were joint R&D, 25% were technological transfers or exchanges, and 42% were OEM supply relationships. Thus, although General Electric was slightly less active than Toshiba during this time period, the proportions of different forms of linkages are roughly the same except in the joint R&D category where General Electric had half as many as Toshiba.

In contrast, the organization set of new linkages in 1984 and 1985 for General Motors is very different. GM formed 11 linkages with 7 Japanese companies: 55% were joint ventures, 18% were joint R&D, 18% were technological transfers or exchanges, and 9% were OEM supply relationships. Honda, on the other hand, has pursued a direct investment strategy as shown in Figure 3.1 rather than a joint venture or technological exchange strategy. Thus, there is a marked contrast in the strategy of a large auto company in the United States and a large auto company in Japan, whereas the strategies of the two organizations in high technology are relatively similar.

Conceptualization of the relation of members of the organization set to the focal organization can also be done using dimensions such as multiplexity and vertical or horizontal interdependence. For example, Figure 3.3 shows that Toshiba has a multiplex relation with Diasonics and General Electric. Its joint ventures are mostly horizontally interdependent, its technological exchanges and joint R&D are backward vertically interdependent, and the OEM supply relations are forward vertically interdependent.

Research Applications of a Network Perspective at Multiple Levels of Analysis

Four key levels of analysis can be synthesized from the major macro theoretical perspectives on interorganizational relationships: (1) the individual level, (2) the organizational level, (3) the population/grouping level, and (4) the community/organizational field level (see Table 3.5). It is important to note that although the general distinctions between levels are clear, the specific boundaries between levels may be ambiguous. At

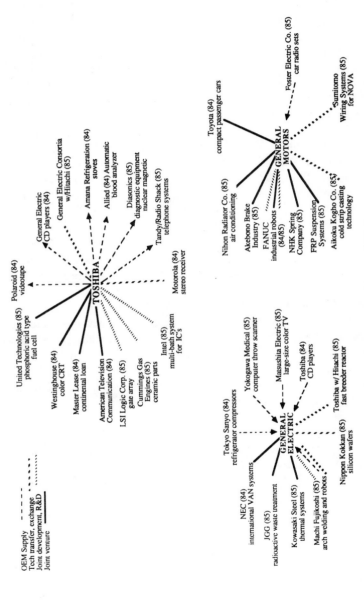

Figure 3.3. Organization Sets of Toshiba, General Electric, and General Motors

NOTE: All linkages were created in 1984 and 1985.
SOURCE: Based on case data from *Cooperations Between American and Japanese Firms: Cases of Industrial Cooperation in 1984* (New York: Japanese External Trade Organization, pp. 1-154); *Cooperations Between American and Japanese Firms: Cases of Industrial Cooperation in 1985* (New York: Japanese External Trade Organization, pp. 1-278); and *JETRO Monitor* (1989).

each level of analysis, a network perspective prompts new research questions.

Individual Level

The individual level of network analysis can be defined as the study of how people affect interorganizational relations and the effects of interorganizational relations on individuals. Questions addressing the former topic might examine the role of boundary spanners in the creation, persistence, and evolution of linkages such as technological transfers. Aldrich (1979) conceptualizes the area, but much more attention should be paid to the relationship of individual ties to interorganizational linkages (Rogers & Kincaid, 1981).

Mizruchi and Stearns (1988) and others have studied many aspects of interlocking directorates, but less attention has been paid to the effects of these types of personal relationships on interorganizational linkages. Mapping networks of interlocking boards of directors with networks of interorganizational linkages might be a first step. Also of interest is how boundary-spanner characteristics (demographics, networks, functional backgrounds, and experiences) are related to the creation of new forms, to the life cycles of linkages, and to the diffusion of innovation in interorganizational networks.

A related topic would be factors to consider in selecting boundary spanners and managers to oversee interorganizational relations such as technological transfers, and the kinds of characteristics of individuals associated with success in managing different forms. The impact of upper-echelon attitudes on the creation and evolution of linkages and networks would be another avenue of research.

The impact of interorganizational relations on individuals is the sixth research area in which little work has been done. How a firm's interorganizational set and networks affect upper-echelon attitudes, decision making, individual power, and career advancement would be one interesting set of questions. The impact of interorganizational linkages such as joint ventures on lower-level employees and research exploring when in a career path a boundary-spanning role is most beneficial also would be important.

TABLE 3.5 Levels of Analysis Across Four Perspectives

	Individual	Organizational	Population/Grouping			Community/ Organizational Field
		firm	*market segment*	*strategic group*	*industry*	"public/private cooperation in R&D consortia and Technopolis (Smilor, Kozmetsky & Gibson, 1988)
Strategy		business unit corporation	"consists of buyers who seek the same offering" (O'Shaughnessy, 1988, p. 107)	"firms that embrace the same *strategies* for serving customers, their competitive postures are similar" (Harrigan, 1985a, p. 13)	"a market in which similar or closely related products are sold to buyers" (Porter, 1985, p. 233)	
		organization	*industry*			
Transaction Cost		"governance structure" (Williamson, 1985, p. 13)	boundaries not clearly defined			

(continued)

Table 3.5. Continued

	Individual	Organizational	Population/Grouping	Community/Organizational Field
	individual level	*organization*	*industry*	*community*
Resource Dependence	"interlocking boards of directors"	"a coalition of groups and interests, each attempting to obtain something from the collectivity by interacting with others and each with its own goals and preferences" (Pfeffer and Salancik, 1978, p. 36)	often defined as 2 digit SIC code level	"functionally integrated systems of interacting populations" (Astley 1985, p. 234)
	individual level	*organization*	*population*	*species*
Ecology	"boundary spanners"	"goal directed boundary maintaining activity systems" (Aldrich, 1979, p. 4)	"a group of organizations that are similar in the competence needed to produce a product or service that is essential to their continued survival" (McKelvey, 1982, p. 24)	"polythetic groups of competence sharing populations isolated from each other because their dominant competencies are not easily shared or transmitted" (McKelvey, 1982, p. 192)
				"set of organizations oriented toward some collective end" (Aldrich & Marsden, 1988, p. 383)

Organizational Level

This level focuses on organizational characteristics and their relationship to the creation, management, maintenance, persistence, and failure of interorganizational relations. Rather than viewing the organization as an isolated atom, a firm is analyzed within its context of relations. Research is beginning to emerge that reflects this perspective, but more is needed to establish generalizability and reliability.

Strategic questions include re-evaluating topics such as organizational motives for creating linkages; how to choose a partner; how to negotiate a linkage; and competitive trade-offs of different forms of linkages, taking into consideration the webs of linkages the organization and its competitors are engaged in currently. Research on what types of portfolios of linkages are most effective in what types of environments would be extremely useful.

Case studies are critical of topics such as: interorganizational negotiation processes (see Weiss, 1987); the transformation of different forms of interorganizational relations over time; longitudinal analyses of the changes an organization set undergoes over time; and the impact of linkage characteristics, their nature, and number on organizational evolution. For example, a number of Japanese companies have formed joint ventures with U.S. companies that evolved to 100% Japanese owned within a few years. Longitudinal analyses can track such patterns, and case studies can help us to understand the dynamics underlying them.

A third area would be the impact of organizational characteristics such as size, age, technology, and structure on the creation and transformation of organization sets and linkages. The relationship of linkage formation to certain stages of organizational life cycles would be especially interesting. For example, do linkages provide a means of overcoming the liabilities of age and size (Aldrich and Auster, 1986)? Can a large, aging organization use joint ventures with small, innovative companies to revitalize? Can two large, aging organizations use a joint R&D linkage to foster creativity in inertia-laden systems?

How does the nature and distribution of ties within a network or organization affect an organization's access to resources, organizational power and dependencies, and its ability to position itself optimally to change with environmental conditions? For example, in July 1987, the U.S. Senate voted to ban imports of Toshiba products to punish the company for its subsidiary having sold superquiet submarine propellers to the Russians ("A Leak That Could Sink the U.S. Lead in Submarines," 1987; "Congress Wants Toshiba's Blood," 1987). Had the government been aware of Toshiba's organization set and the extent of Toshiba's interdependencies with U.S. companies, it might have anticipated and managed more effectively the backlash and lobbying that occurred from Toshiba's U.S. partners.

Organizational Population/Groupings Level

Population is used here as a generic term that refers to various groupings of organizations used across strategic, transaction-cost, resource-dependence, and ecological perspectives. Distinctive shared competencies within a specified time interval (Beard & Dess, 1988) would be the defining characteristic binding these groupings and distinguishing them from higher and lower levels of analysis. *Competencies* are defined as the set of technical, managerial, and operational knowledge and skills needed to produce the primary product or service (Beard & Dess, 1988; McKelvey, 1982).

Within this category, strategic groups and markets typically would be based on more narrowly defined shared competencies, whereas species and industries may cut across strategic groups and market segments and would be based on broader definitions of shared competencies. Furthermore, strategic groups, species, populations, and industries are based on aggregations of organizations, whereas market segment refers more to characteristics of customers.

Research areas at this level of analysis include basic and comprehensive information of the range of different forms of linkages, the composition of organization sets, and the structure of networks within a grouping/population. Beyond descriptive background, it would be useful to learn more about the dynamics of exchange relations resulting from linkages and networks and how they alter power and dependencies within a population. An extension might be explicit analyses of the extent of dependence of U.S. industries on key Japanese companies and competitors.

How are environmental and industry characteristics related to the creation, persistence, and decline of different forms of linkages, and the development of different types of organization sets and networks over time? For example, based on Figures 3.1, 3.2, and 3.3, one might question whether the environments of high-technology industries tend to make low-commitment linkages such as technological transfers and exchange, joint R&D, and OEM supply linkages more attractive, whereas the environments of more mature industries such as steel or automobiles make joint ventures more attractive. More fundamentally, a network orientation may challenge our view of the environment. Rather than a set of static attributes that organizations respond to, we may begin to see environments as flows of resources and transactions not demarcated clearly from the organizations immersed within them (Pfeffer, 1987).

Jorde and Teece (1989) recently argued that U.S. antitrust laws currently facilitate overseas linkages rather than domestic linkages. They raise the question of whether laws that make it easier for a U.S. automaker to hook up with a Korean or Japanese automaker than to hook up with a domestic competitor are healthy for the U.S. economy. As more network ties are established, anticipating the reaction of these collectivities to proposed legal changes and understanding the effects of legal changes on those webs will become increasingly critical.

Community/Organizational Field

The community or organizational field is a functionally integrated system of interacting populations (Aldrich & Marsden, 1988; Astley, 1985; Carroll, 1984; Warren, 1967). Community-level linkages and networks differ from population-level forms because the focus is on phenomena such as consortia, trade associations, or other strategic alliances that cut across multiple populations and institutions to achieve common or mutually supportive interests. For example, the formation of worldwide technopoleis or high-tech centers is motivated, in part, by interorganizational networking across academic, business, and government organizations (Smilor, Kozmetsky, & Gibson, 1988).

Community-level forms of networks such as the *kinyu keiretsu* (financial linkages) or *kihyo shedan* (enterprised group) have existed within Japan for decades (Gerlach, 1987), and in recent years have sprung up between the United States and Japan and within the United States itself. The Boeing, Mitsubishi, Fuji, and Japanese government consortium which is building and improving the Boeing 767 (Roehl & Truitt, 1987) or the Texas Instruments, Motorola, and Hitachi consortium developing next-generation chips would be examples of international consortia between the United States and Japan.

Within the United States, domestic consortia have emerged largely due to the passage of the National Cooperative Research Act in 1984 and increasing competitive pressure from Japan. The Microelectronics and Computer Technology Corporation (MCC) and Semiconductor Manufacturing Technology (SEMATECH), both located in Austin, Texas, are two of the most well-known consortia in the United States. MCC research focuses on advanced computing technology, semiconductor packaging/interconnect, software technology, computer-aided design, and superconductivity, has 21 member companies, and has an annual research budget of $65 million (Gibson & Rogers, 1988). SEMATECH is comprised of 14 U.S. semiconductor companies and was formed in 1987 in response to the devastation of the dynamic random-access memory market in the United States by the Japanese. These U.S. consortia are intriguing because they transcend typical competitive boundaries in pursuit of a collective goal of regaining market share. Understanding how to manage these forms effectively and their implications for both national and global economies will become issues that must be addressed as these forms proliferate.

Conclusions

Previous research on interorganizational relationships has focused primarily on dyads or triads at the organizational level of analysis. A network perspective broadens this orientation, offering new perspectives and raising new research questions. The empirical challenges of this type of research are great. Comprehensive, longitudinal data on large samples of linkages, organization sets, or networks are required. Such a network perspective is necessary, however, for U.S. public and private institutions to acquire a more sophisticated approach to international interorganizational relationships so that they will be able to compete successfully in the global economy.

Notes

1. A few researchers have used network analysis to study international interorganizational relationships. See, for example, Jarillo (1988), Thorelli (1986), or Walker (1988).

2. The case information was obtained through the Japanese External Trade Organization and is supplemented by qualitative interviews with top managers in the United States and Japan, March 1989.

References

A leak that could sink the U.S. lead in submarines. (1987, May 18). *Business Week,* pp. 65-66.

Aldrich, H. (1979). *Organizations and environments.* Englewood Cliffs, NJ: Prentice Hall.

Aldrich, H., & Auster, E. (1986). Even dwarfs started small: Liabilities and age and size and their strategic implications. *Research in Organizational Behavior, 8,* 165-198.

Aldrich, H., & Marsden, P. (1988). Environments and organizations. In N. Smelser, (Ed.), *Handbook of sociology* (pp. 361-392). Newbury Park, CA: Sage.

Aldrich, H., & Whetten, D. (1981). Organization-sets, action-sets, and networks: Making the most of simplicity. In P. C. Nystrom & W. Starbuch (Eds.), *Handbook of organizational design* (Vol. 1, pp. 385-408). London: Oxford University Press.

Astley, W. G. (1985). The two ecologies: Population and community perspectives on organizational evolution. *Administrative Science Quarterly, 30,* 224-241.

Auster, E. R. (1990). *The relationship of industry evolution to patterns of technological linkages, joint ventures, and direct investment between U.S. and Japan.* Academy of Management Proceedings, San Francisco, CA.

Auster, E. R. (1987). International corporate linkages: Dynamic forms in changing environments. *Columbia Journal of World Business, 22*(2), 3-6.

Barnes, J. A. (1972). *Social networks.* Reading, MA: Addison-Wesley.

Beard, D., & Dess, G. (1988). Modeling organizational species interdependence in an ecological community: An input-output approach. *Academy of Management Review,* 13, 362-373.

Berlew, F. (1984, July-August). The joint ventures: A way into foreign markets. *Harvard Business Review,* 48-52.

Bivens, K. K., & Lovell. E. (1966). *Joint ventures with foreign partners.* National Conference Board, New York.

Blau, P. (1964). *Exchange and power in social life.* New York: Wiley.

Brahm, R., & Astley, G. (1988). *Constrained exploitation: A re-evaluation of governance and performance in joint ventures.* Paper presented at the meeting of the Academy of Management, Anaheim, CA.

Buckley, P., & Casson, M. (1988). A theory of cooperation in international business. In F. Contractor & P. Lorange (Eds.), *Cooperative strategies in international business* (pp. 31-54). Lexington, MA: Lexington.

Burt, R. (1983). Models of network structure. *Annual Review of Sociology, 6,* 79-141.

Burt, R. (1988). The stability of American markets. *American Journal of Sociology,* 356-395.

Burt, R., Minor, M., & Associates (1983). *Applied network analysis: A methodological introduction.* Beverly Hills: Sage.

Business without borders. (1988, June 20). *U.S. News and World Report,* pp. 48-53.

Carroll, G. (1984). Organizational ecology. *Annual Review of Sociology, 10,* 71-93.

Congress wants Toshiba's blood. (1987, July 6). *Business Week,* pp. 46-47.

Contractor, F., & Lorange, P. (1988). *Cooperative strategies in international business.* Lexington, MA: Lexington.

Cook, K. (1977). Exchange and power in networks of interorganizational relations. *Sociological Quarterly,* 18, 62-82.

Cook, K. (1982). Network structures from exchange perspective. In P. Marsden & N. Lin (Eds.), *Social structure and networks analysis* (pp. 177-199).

Doz, Y. (1988). Technology partnerships between larger and smaller firms: Some critical issues. In F. Contractor & P. Lorange, (Eds.) *Cooperative strategies in international business* (pp. 317-338). Lexington, MA: Lexington.

Evan, W. (1972). An organization set model of interorganizational relations. In M. Tuite, R. Chisholm, & M. Radnor (Eds.), *Interorganizational decision making* (pp. 181-200). Chicago: Aldine.

Evan, W. (1963). The organization set: Toward a theory of interorganizational relations. *Management Science, 11,* 217-230.

Gerlach, M. (1987). Business alliances and the strategy of the Japanese firm. In G. Carroll & D. Vogel (Eds.), *Organizational approaches to business strategy* (pp. 27-143). Cambridge, MA: Ballinger.

Gibson, D., & Rogers, E. (1988). The MCC comes to Texas. In F. Williams (Ed.), *Measuring the information society.* Newbury Park, CA: Sage.

Harrigan, K. R. (1987). Strategic alliances: Their new role in global competition. *Columbia Journal of World Business, 22*(2), 67-70.

Harrigan, K. R. (1985a). *Strategic flexibility.* Lexington, MA: Lexington.

Harrigan, K. R. (1985b). *Strategies for joint ventures.* Lexington, MA: Lexington.

Hawley, A. (1950). *Human ecology.* New York: Ronald Press.

Hergert, M., & Morris, D. (1987). Trends in international collaborative agreements. *Columbia Journal of World Business, 22,* 15-21.

Hirsch, P. (1972). Processing fads and fashions: An organization-set analysis of cultural industry systems. *American Journal of Sociology, 77*(4), 639-59.

Hladik, K. (1988). R&D and international joint ventures. In F. Contractor & P. Lorange (Eds.), *Cooperative strategies in international business* (pp. 187-204). Lexington, MA: Lexington.

Homans, G. (1961). *Social behavior.* New York: Harcourt, Brace and World.

Jarillo, J. C. (1988). On strategic networks. *Strategic Management Journal, 9,* 31-41.

JETRO Monitor. (1989, January). Newsletter on Japanese Economics and Trade Issues, *3*(9), 1-4.

Jorde, T., & Teece, D. (1989, June 18). Antitrust law's drag on innovation. *The Wall Street Journal,* p. A17.

Killing, J. P. (1983). *Strategies for joint venture success.* New York: Praeger.

Kogut, B., & Singh, H. (1988). Entering the United States by joint venture: Competitive rivalry and industry structure. In F. Contractor & P. Lorange (Eds.), *Cooperative strategies in international business* (pp. 241-253). Lexington, MA: Lexington.

Laumann, E., Marsden, P., & Prensky, D. (1983). The boundary specification problem in network analysis. In R. Burt, M. Minor, & Associates (Eds.), *Applied network analysis* (pp. 18-34). Beverly Hills, CA: Sage.

McKelvey, B. (1982). *Organizational systematics: Taxonomy, evolution, and classification.* Berkeley: University of California Press.

Mizruchi, M., & Brewster-Stearns, L. (1988). A longitudinal study of the formation of interlocking directorates. *Administrative Science Quarterly, 33*(2), 194-210.

O'Shaughnessy, J. (1988). *Competitive marketing: A strategic approach.* Boston: Unwin Hyman.

Pennings, J. (1981). Strategically interdependent organizations. In P. Nystrom & W. Starbuck (Eds.), *Handbook of organizational design* (Vol. 1, pp. 433-455). London: Oxford University Press.

Pfeffer, J. (1987). Bringing the environment back in: The social context of business strategy. In D. Teece (Ed.), *The competitive challenge: Strategies for industrial innovation and renewal* (pp. 117-135). Cambridge, MA: Ballinger.

Pfeffer, J., & Salancik, G. (1978). *The external control of organizations: A resource dependence perspective.* New York: Harper & Row.

Pisano, G., Russo, M., & Teece, D. (1988). Joint ventures and collaborative arrangements in the telecommunications equipment industry. In D. Mowery (Ed.), *International collaborative ventures in U.S. manufacturing* (pp. 23-70). Cambridge, MA: Ballinger.

Pisano, G., Shan, W., & Teece, D. (1988). Joint ventures and collaboration in the biotechnology industry. In D. Mowery (Ed.), *International collaborative ventures in U.S. manufacturing* (pp. 183-222). Cambridge, MA: Ballinger.

Porter, M. (1985). *Competitive advantage.* New York: Free Press.

Pucik, V. (1988). Strategic alliances with the Japanese. In F. Contractor & P. Lorange (Eds.), *Cooperative strategies in international business* (pp. 487-498). Lexington, MA: Lexington.

Roehl, T., & Truitt, J. F. (1987). Stormy, open marriages are better: Evidence from U.S., Japanese, and French cooperative ventures in commercial aircraft. *Columbia Journal of World Business, 21*(2), 87-96.

Rogers, E., & Kincaid, D. L. (1981). *Communication networks: Toward a new paradigm for research.* New York: Free Press.

Smilor, R., Kozmetsky, G., and Gibson, D. (1988). *Creating the technopolis: Linking technology commercialization and economic development.* Cambridge, MA: Ballinger.

Sanger, D. (1989, January 29). For American technology, it's a job just to keep within sight of the Japanese. *The New York Times*, pp. D1, D3.

Thorelli, H. (1986). Networks: Between markets and hierarchies. *Strategic Management Journal, 7,* 37-51.

Tichy, N. (1981). Networks in organizations. In P. Nystrom & W. Starbuch (Eds.), *Handbook of organization design.* London: Oxford University Press.

Tichy, N., Tushman, M., & Fombrun, C. (1979). Social network analysis for organizations. *Academy of Management Review, 4,* 507-519.

Walker, G. (1988). Network analysis for cooperative interfirm relationships. In F. Contractor & P. Lorange (Eds.) *Cooperative strategies in international business* (pp. 227-240). Lexington, MA: Lexington.

Warren, R. (1967). The interorganizational field as a focus for investigation. *Administrative Science Quarterly, 12,* 396-419.

Weiss, S. (1987). Creating the GM-Toyota joint venture: A case in complex negotiation. *Columbia Journal of World Business, 22*(2), 23-37.

Williamson, O. (1985). The economic institutions of capitalism. New York: Free Press.

III

Contexts of
Technology Transfer

4

Research Consortia:
The Microelectronics and
Computer Technology Corporation

CHRISTOPHER M. AVERY
RAYMOND W. SMILOR

Consortia represent a new way for U.S. industries to conduct R&D. They allow companies within and across industries to cooperate within one domain while competing in others. The cooperation nurtures the joint development of emerging technology, yet the firms continue to compete in the ways they take products and services to the marketplace. The Microelectronics and Computer Technology Corporation (MCC) is a prime example of an R&D consortium charged directly with technology transfer. In this chapter, researchers Christopher M. Avery and Raymond W. Smilor report their study of technology transfer processes within MCC. Avery is employed by Technology Futures, Inc., an innovation management firm. The materials in this chapter are based in part upon his doctoral dissertation completed at the University of Texas at Austin (Avery, 1989). Smilor, known for his research

into the development of high technology corporations and tech-
nological cities, has been directing a large-scale study of research
consortia of which this chapter is also a part. Smilor is executive
director of the IC² Institute at the University of Texas at Austin.

A New Organizational Form

Late in 1982, William Norris, CEO of Control Data Corporation, called together the chiefs of 10 of the United States' largest computer corporations.[1] He also invited the U.S. Department of Justice to the meeting. Why? He planned to ask the computer corporation chiefs to do something that would challenge antitrust laws. Norris predicted that if these 10 companies did not cooperate in some way, within 20 years there would be only three players in the computer industry—AT&T, IBM, and Japan, Inc. (Gibson & Rogers, 1988). He proposed that the corporations collaborate in the development of technology, and that the Department of Justice sanction their collaboration. In 1983, the 10 corporations formed the Microelectronics and Computer Technology Corporation (MCC), a for-profit technology research corporation owned and controlled by the founding firms. By 1984, the National Cooperative Research Act was passed clarifying the antitrust laws with regard to R&D cooperation, effectively giving MCC a green light to proceed to collaborate on joint research and development activities. Through July 1989, the Department of Justice had received 135 filings for the formation of new R&D consortia. The rapid growth of these types of collaborative R&D organizations is a sign of the member firms' desire to cooperate nationally so as to compete globally.

Cooperating to Compete

Consortia represent a new way for U.S. industries to conduct R&D. In essence, the member firms collaborate on the precompetitive development of technology. Once that technology research reaches the point of "proof of concept," the consortium ceases to develop it and turns it over to the member companies who pursue specific applications and market opportunities. Thus, technology transfer is key to the success of an R&D consortium.

Consortia take many forms. Some employ no researchers directly, choosing instead to distribute research funds to university scientists who then share their research reports among the consortia members. Other consortia, like MCC and SEMATECH, operate with large budgets and their own buildings, employees, and goals for profitability. At MCC, organizational members provide some researchers and staff, while others are hired directly by MCC.

Organizational collaborations are not new, but the closeness of collaborations in these R&D consortia is (Dimancescu & Botkin, 1986). Industry consortia pool the members' research funds, thereby allowing member firms to (1) leverage their investment in research and development; (2) reduce the amount of duplicated research effort in an industry; (3) promote the long-term focus of basic research; and (4) reduce the risk of failure by diversifying the portfolio of research projects. Hence, the international competitiveness of an industry is likely to be increased (Petit, 1987).

Development of U.S. R&D Consortia

The 135 filings for new R&D consortia that had been approved by the Department of Justice through July 1989, are comprised of 1149 entities, distributed as follows (IC^2 Institute, 1989): 83% (959) are U.S. companies; 8% (90) are foreign members; 4% (40) are associations, institutes, or councils; 2% (28) are consortia that are members of other consortia; 1% (13) are U.S. universities; 1% (10) are federal government departments; and 1% (9) are state governments. These 135 consortia are distributed across a diverse set of industries (e.g., telecommunications, automotive, environmental, microelectronics and computing, energy, materials, building/construction, chemicals, manufacturing, biotechnology, health care, electronics, heavy machinery, and intellectual property), and all share the major objective of facilitating technological innovation. In addition, many large U.S. companies belong to more than one consortium. Bellcore—itself a consortium—belongs to 22 other consortia. In all, 16 U.S. companies each belong to six or more R&D consortia.[2]

Three characteristics of consortia make technology transfer an especially challenging issue: (1) the interaction of multiple constituent cultures within one organization; (2) the simultaneous compete-and-cooperate arrangement of the member organizations; and (3) the complex nature of traditional technology transfer processes. A case study of technology transfer processes within MCC offers insights into the communication perspectives on these three characteristics.

The Study and Setting

Because member firms need to speed the development of technology and to move it to the marketplace, technology transfer is a primary mission of consortia and a major criterion of their success. By virtue of multiple shareholders supporting multiple projects, consortia offer a concentration of many possible technology transfer approaches. The Microelectronics and Computer Technology Corporation (MCC) provides an important perspective on technology transfer not only because it was the first major U.S. consortium to register under the National Cooperative Research Act of 1984, but also because it has been the most visible (Peck, 1986). As of 1990, MCC was 7 years into research programs with original delivery horizons in the 6- to 10-year range. Thus, the transfer of research results from MCC to member companies is a useful case to understand better the complex organizational communication task associated with technology transfer. Specifically, our research explored organizational communication processes within and between the consortium and its members. The processes of interest in this study included any component of organizational communication between and among shareholders and the consortium that would affect the use of technology research (Smilor, Gibson, & Avery, 1989).

In discovering and exploring the organizing mechanisms designed to move research into and through the consortium and across its boundaries to the shareholder companies, three key research questions were posed:

(1) What are the mechanisms by which MCC conducts technology transfer?
(2) What organizational communication processes, other than technology transfer mechanisms, influence technology transfer between MCC and its shareholders?
(3) How do the technology transfer mechanisms and other organizational communication processes relate to one another?

Like many of the other R&D consortia in the United States, MCC serves to pool some of the R&D funds of member companies in order to reduce duplication of effort and to spread the associated risk. Unlike many other consortia, MCC does not contract primarily with universities to perform research, but instead employs its own staff of about 450 and is housed in its own 200,000 square-foot facility (as well as additional buildings). By 1987, MCC was funded at about $70 million per year by its 19 medium- to large-sized member corporations. Its function is twofold: (1) conduct precompetitive research of importance to its member companies, and (2) transfer the results to member companies. MCC is headquartered at the University of Texas at Austin Balcones Research Center on the northern outskirts of Austin, Texas.

Program Descriptions and Examples of MCC Technology

Packaging/interconnect. This research program focuses on advanced chip packaging technologies and methods for connecting faster and highly integrated chips. A core group and three research satellites—bonding and assembly development, multichip system technology, and interconnect technology—comprise the packaging/interconnect program. An example of a technology transfer from packaging/interconnect is improvements to tape-automated bonding. The tape is supplied to MCC by 3M, one of the consortium's member companies. There has been support for MCC's helping 3M improve its technology because it raises the standard of the industry.

Advanced computer technology. The advanced computer technology program focuses on (1) expert systems and large-scale knowledge bases, (2) human interface tools, and (3) advanced languages and object-oriented data bases. Research takes place in a core operation as well as three satellites: artificial intelligence, human interface, and systems technology. An example of a technology transferred from the advanced computer architecture program is Proteus, an expert systems environment incorporating forward and backward chaining of rules, truth maintenance, and an object-oriented data base. NCR Corporation has used Proteus to build an expert system to aid in chip design.

Software technology. This program focuses on tools that will improve significantly the productivity and quality of software development for the shareholders. The research focuses specifically on the early stages of software development where design specifications may not be articulated clearly, yet complex networks of teams of designers must participate in the design process. An example of a technology transferred from the software technology program to a number of companies is GIBIS (graphical issue-based information system). GIBIS is a technology for recording and displaying the evolution of issues and arguments made in group design meetings.

Computer-aided design. The computer-aided design program's near-term goal is to develop an advanced software environment for CAD tool development. The long-term goal involves designing better circuitry and systems for CAD processing, eventually supporting the design of systems containing 50 million devices.

High-temperature superconductivity. The superconductivity program investigates new electronic applications for high-temperature superconductors. The program is looking at hybrid systems, fabricating materials, and thin-film applications.

Research Method

Due to the lack of theory in technology transfer to guide organizational communication research and the absence of organizational communication research on R&D consortia, the "grounded theory" (Glaser & Strauss, 1967) method was selected for its strength in generating—rather than testing—theory from social science data. The data were provided by MCC archives, transcripts from semistructured interviews, and manuscripts of field observation notes. The grounded theory approach allows the researcher to construct empirical generalizations (as opposed to an anecdotal account) through an inductive analysis of such qualitative case study data (Martin & Turner, 1986). Finally, grounded theory is appropriate for the general movement in the field of organizational communication toward more descriptive and inductive accounts of complex and contextual behavior in organizations (Alderfer & Smith, 1982).

Interview transcripts and field note manuscripts were divided into 810 "incidents"; an incident is the smallest unit of data that still contains a complete thought. Each incident was compared to each other incident and grouped into 22 categories, which in turn were grouped into four clusters. These clusters and categories were analyzed in order to answer the research questions. The list of clusters and categories is given in Table 4.1.

TABLE 4.1: Clusters and Their Categories

CLUSTERS	CATEGORIES

The Consortium
  Organizational Structure
  Program Description
  Description of a Technology
  Shareholder Description
	MCC Setting

Conducting Technology Transfer
  Bridge
  Transfer Axis
  Evaluating Success
  Incentive and Resistance
  Assessing Utilization
	Transfer Mechanisms

Technology Differences
  Equivocality
  Continuity
	Level

Integration
  Intra-MCC Communication
  Intrashareholder Communication
  Intershareholder Communication
  Black Box
  Shareholder/MCC Interdependence
  Links between Internal and Interorganizational
    Communication
  Public Relations
	Alignment

The first cluster, the *consortium*, describes the organizing characteristics and contextual information concerning MCC. The second cluster, *conducting technology transfer,* reveals the surface structure of the technology transfer process. Many answers to the first research question were found in this cluster. *Technology differences,* the third cluster, is a core cluster in that it revealed itself early in the study and was confirmed and strengthened throughout the analyses. This cluster reveals a powerful and important substructure of the technology transfer process. It is related to all other clusters and is important in answering the second and third research questions. The final cluster, *integration,* captures elements of communication that influence the integration of MCC and shareholders.

Analyses of Technology Transfer Mechanisms

MCC employs 20 different technology transfer mechanisms, as listed in Table 4.2. Mechanisms are grouped according to communication mode. Five mechanisms are passive, 15 are active. Passive mechanisms appear to have two functions: Mechanisms such as technical reports and journal papers act as recorders and accumulators of knowledge upon which others then build. The second function served by these passive mechanisms is to create awareness through newsletters and videos. Active mechanisms function much differently in that they focus on the social processes of preparing for, planning for, sustaining, or buffering the effects of technology transfer. They also aid in transporting technical information from one location to another. The shareholder representative (a shareholder employee who resides at MCC for an extended period of time and acts as a transfer agent to the member company) and the shareholder assignee (like the shareholder representative without the technology transfer responsibilities) are examples of active mechanisms of technology transfer.

There are two types of technology transfer mechanisms that are not designed to communicate research results, but to actively prepare and orient participants for future transfer possibilities. These anticipatory technology transfer processes appear in at least two forms. In the first form, MCC shareholder meetings and shareholder committees involve shareholders in research planning long before results are ready for transfer. This activity begins the process of linking MCC research strategy with member company product-development strategies. Members can begin to anticipate how future research results will be incorporated into member company products or processes. In the second form, MCC encourages shareholders to invest internally in research programs that compete with, track, receive, or measure MCC research results. These include the receptor organization (a shareholder group with responsibility for receiving the technology) and shadow research projects (shareholder research in the same technology that MCC is researching), as well as collaborative activities. These mechanisms greatly affect the member organizations' desire and ability to move MCC technology across their boundaries.

Another active transfer mechanism employed by MCC is transferring technology to the shareholder companies by transferring people. This approach has proven to be complicated to manage for the consortium

TABLE 4.2: Technology Transfer Mechanisms at MCC

Passive Technology Transfer
  Proprietary Technical Reports
  Refereed Journal Articles
  Newsletters
  Videotaped Overviews
  Videotaped Demonstrations

Active Technology Transfer
  Meetings of MCC and Shareholders
    Software Technology Advisory Council (STAC)
    Technical Advisory Board (TAB)
    Program Technical Advisory Board (PTAB)
    Slide Presentations
  Shareholder Committees
    Program Advisory Committee (PAC)
    Technical Requirements Panel
    Manufacturability Panel
    Quality Assurance Panel
  Shareholder Representatives
  Shareholder Assignees
  Visitors Program
  Shareholder Site Demonstrations
  Receptor Organizations within Shareholders
  Shadow Research Projects within Shareholders
  Shareholder/MCC Collaboration

and shareholder organization. Originally, shareholder employees were to spend a number of years at MCC, making frequent trips to their shareholder companies. A number of problems emerged, however, when it came time to return the transfer agents to the shareholders: First, shareholder representatives often did not retain their once-held status at the shareholder companies, which contributed to resistance in attempting to transfer technology. Second, because of the attractiveness of the consortium environment, shareholder representatives often did not want to leave MCC. Third, researchers seldom chose to transfer out of research groups into development groups; instead, they preferred to continue to work at basic rather than applied levels of technology research. Fourth, in the consortium and shareholder arrangement, the management of social processes between competing organizations was a more important task for the transfer agent than simply acquiring and carrying the research knowledge. MCC has learned that transfer agents should be stationed close to the shareholders—as well as close to MCC and the research—so that meaningful contacts can be initiated and maintained within the shareholder companies.

Technology Characteristics

Characteristics of the technology influence how transfer mechanisms are designed, selected, and employed, and to what extent they are successful. Three technology variables were found to affect communication in technology transfer.

(1) *Level,* the degree to which the work on a technology is basic as compared to applied;

(2) *Equivocality,* the degree to which interpretations of a technology vary across technologies and across cultures; and

(3) *Continuity,* the degree to which communication in technology transfer is relatively discrete (one-way and of short duration) as opposed to relatively continuous (interactive or transactive and sustained).

The first dimension, *technology level*, has been a component of previous research on technology innovation (Bright, 1964), but has not been related previously to organizational communication. The second dimension, *technology equivocality*, is similar to message equivocality in communication research (Daft & Macintosh, 1981), but also has not been identified previously in studies of technology transfer. Characteristics of the third dimension, *technology continuity*, have been identified previously in technology transfer research, but have not been linked with the other contextual factors as they are here. Those characteristics include (1) passive versus active mechanisms; (2) interpersonal versus mediated communication modes; and (3) transportation (one-way) as opposed to transactive (two-way) models of technology transfer.

The technology-level dimension accounts for a large number of findings concerning organizational communication and technology transfer, and accounts for much of the variance attributed to success or failure of technology transfer. Transfer mechanisms symbolically convey the level of technology and its importance. For instance, technology will be perceived as close to or far from commercialization depending upon whether it is communicated through research reports (far from commercialization) or demonstrations (close to commercialization).

The equivocality of a technology varies across both technologies and shareholders. In the first instance, technologies that are highly equivocal—those that have many possible interpretations or applications—are perceived as more difficult to transfer and are associated with a search for technology transfer mechanisms capable of dealing with the variety of possible interpretations. Therefore, MCC may customize technology transfer attempts based on unique perceptions of the technology by the shareholder. In the second instance, where equivocality varies across shareholders, the communication relationships between MCC and shareholders vary more for equivocal than for unequivocal technologies. Hence, ways in which technologies vary (and MCC works on hundreds of technologies) and ways in which shareholder cultures vary (and MCC transfers to 19 company cultures) create tremendous variety for the technology transfer mechanisms.

The continuity dimension integrates and extends a number of previous findings concerning organizational communication and technology transfer. Previous studies have identified transfer mechanisms as passive or active, and as interpersonal or mediated. Furthermore, many others have argued that technology transfer is a transactive process as opposed to a process of transporting a technology across organizational boundaries. This dimension shows that transfer mechanisms can be located along a continuum where at one end are relatively discrete processes and at the other end are relatively continuous processes. This finding joins and reconciles the previously incompatible assumptions of Rogers' diffusion model (1983) and Sawhney's creativity model (Williams et al., 1988) of technology transfer. The continuity dimension can explain both one-way and two-way communication processes.

Technology Transfer Mechanisms and Other Communication Processes

Relationships between organizational communication and technology transfer processes were identified by comparing shareholders (strategy, size, internal R&D investment, and rate of change), technologies (level, equivocality, and continuity), and MCC research programs (strategy, structure, and technology). In general, the combination of variables induces tremendous variety for MCC in managing technology transfer. MCC has responded to this variety by increasing integration across organizational boundaries in order to strengthen cooperation, and by increasing variety by experimenting with and customizing technology transfer mechanisms.

Across shareholders, investment in MCC does not translate into the same uses or benefits of technology transfer. Large shareholders with internal R&D investments in areas similar to MCC research tend to: (1) engage in a greater variety of contacts with MCC (i.e., different people interact with MCC in more ways); (2) prefer MCC to conduct basic—that is, long-term—research; (3) prefer to receive research results in a continuous and informative manner; and (4) use MCC information primarily as research intelligence to guide in-house decisions on their own precompetitive research. Relatively small shareholders in terms of size and internal R&D investments tend not to have internal R&D investments in areas similar to MCC. They (1) engage in more stable relationships with MCC (few people visiting the same MCC people more often); (2) prefer MCC to conduct applied research; (3) prefer discrete transfers of applications; and (4) use MCC information more in terms of product development.

In comparing the continuity dimension across technologies and shareholders, it was found that continuity varies with both level and equivocality of technology. Although level and equivocality often may co-vary, they are seen as orthogonal dimensions. This finding hypothesizes that level and equivocality can be viewed as perpendicular axes with continuity varying directly with each (see Figure 4.1). The model predicts that an objective and applied technology can be transferred successfully through discrete transfer processes, such as in the diffusion-of-innovations process popularized by Rogers (1983). These processes tend to transfer a finished technology from an expert environment to a nonexpert environment using target marketing and persuasion techniques. On the other hand, the model predicts that a subjective and basic technology is suited best for a continuous transfer process inviting creativity and collaboration. These processes tend to transfer conceptual information between entities interested in advancing the concept toward some utility.

The integration of MCC and shareholders over time is associated with evolutionary communication processes. First, funding decisions for MCC have been pushed down within the shareholder organizations. This has increased the number and applied nature of technology transfer requests to MCC. Second, researchers and shareholders often

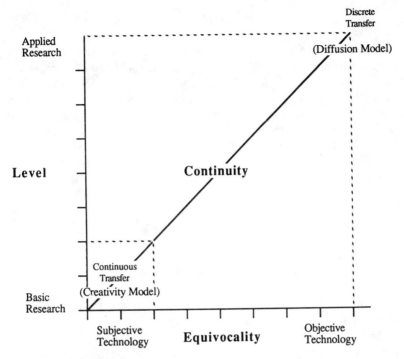

Figure 4.1. Transfer Continuity as a Function of Technology Level and
          Equivocality

SOURCE: Avery (1989).

withhold requested information about work in progress; this is because the researchers may not wish to explain emergent and changing
technologies. After transferring a proof of concept to a shareholder, the
shareholder may resist attempts from MCC to seek feedback information because this would be giving away competitive information. In
response to such challenges, one MCC research program increased
interorganizational ties with its shareholders in order to increase shared
expectations. This was accomplished through convening a series of
technology committees, comprised of members from each shareholder,
to achieve consensus and deliver these technology requirements to
MCC's research program director.

Conclusion

A number of factors combine to make R&D consortia especially intriguing places to view technology transfer: (1) the interaction of multiple-constituent cultures within one organization; (2) the simultaneous competition and cooperation of members; and (3) the nonlinear nature of the technology transfer process. Technology differences (level equivocality and continuity) show that technology transfer is an organizational communication phenomenon affected by the variety of interpretations, messages, symbolic activities, and types of research in which participants engage. The technology transfer process should begin before there are research results (i.e., there is anticipatory transfer which ultimately guides the application of research). Hence, the definition of technology transfer that guided this investigation—the organizing process through which the results of basic and applied research are communicated to potential users—was revised to account for the atemporal and interpretive nature of the process: Technology transfer is the process of integrating science and managerial objectives across entities.

Technology research and transfer proceeds in a disorderly and multidirectional fashion between a consortium and its shareholders. The technology being transferred is seldom a fully formed idea or a finished technology. Therefore, supportive organizational cultures are necessary in order to provide for the continuous and interactive communication required for many transfer situations.

At the interpersonal level, technology transfer can be viewed as a process of convergence (Kincaid, 1979) characterized by mutual causation and interdependence of technology transfer participants over time. From this perspective, communication in technology transfer does not carry the technology; rather, it provides a means for members of different organizations to teach and learn from one another. At the organizational level, technology transfer can be viewed as an interpretive process, and organizations can be viewed as technology interpretation systems (Weick, 1988). From this perspective, communication in technology transfer elicits meaning about technology and organizes behavior around interpretations of technology. At the interorganizational level, technology transfer can be viewed as interaction between an organization (the consortium) and its environment (shareholders).

As Lawrence and Lorsch (1967) discovered in their study of organizations and environments, there is no one best way to organize. Similarly, this study finds that there is no one best way to transfer technology, but that the contingent components in the process can be identified. Managing technology transfer between a consortium and its shareholders requires matching consortium communication structures to those of the shareholders. Finally, at the level of organizational environments, technology transfer can be viewed as an organizational response to turbulent environments (Emery & Trist, 1965). Organizational communication in technology transfer, especially for consortia, forms a web of interorganizational networks that buffers the shareholders from turbulence in their environments and allows them to evolve in ever-increasing nets of interdependence (Terreberry, 1968).

Identifying and modeling the technology transfer process is neither as elusive as previous studies have purported, nor is it as simple as achieving good communication. Technology transfer is affected by an identifiable range of interpersonal, technological, organizational, and environmental variables. As shown by this study, these variables and associated processes can be examined and eventually explained through systematic organizational communication research (Avery & Browning, in press).

Notes

1. The companies were Advanced Micro Devices, Control Data Corporation, Digital Equipment Corporation, Harris Corporation, Honeywell, Inc., Motorola Inc., National Semiconductor, NCR Corporation, RCA, and Sperry. Additions since founding include Bellcore, Boeing Company, Eastman Kodak Company, General Electric Corporation, Hewlett-Packard Company, Hughes Aircraft Company, Lockheed Corporation, Martin Marietta, 3M Company, and Rockwell International. Dropouts include RCA and Sperry.

2. U.S. companies with memberships in six or more consortia (IC^2, 1989) include Bellcore (22), Digital Equipment Company (9), Texas Instruments (8), Rockwell Corporation (8), Hewlett-Packard (8), Ford (8), Honeywell (7), Harris Corporation (7), General Motors (7), Exxon (7), Amoco Corporation (7), Shell Development Company (6), Mobil R&D Corporation (6), International Business Machines Corporation (6), General Electric Company (6), and E. I. du Pont de Nemours (6).

References

Alderfer, C. P., & Smith, K. K. (1982). Studying intergroup relations embedded in organizations. *Administrative Science Quarterly, 27,* 35-65.

Avery, C. M. (1989). *Organizational communication in technology transfer between an R&D consortium and its shareholders: The case of MCC.* Doctoral dissertation, College of Communication, University of Texas at Austin.

Avery, C. M., & Browning, L. D. (in press). *The technology transfer grid: Managing in the decade of cooperation.* Reading, MA: Addison-Wesley.

Bright, J. R. (1964). *Research, development and technological innovation.* Homewood, IL: Irwin.

Daft, R. L., & Macintosh, N. B. (1981). A tentative exploration into the amount and equivocality of information processing in organizational work units. *Administrative Science Quarterly, 26,* 207-224.

Dimancescu, D., & Botkin, J. (1986). *The new alliance: America's R&D consortia.* Cambridge, MA: Ballinger.

Emery, F. E., & Trist, E. L. (1965, February). The causal texture of organization environments. *Human Relations,* 21-31.

Gibson, D. V., & Rogers, E. M. (1988). The MCC comes to Texas. In F. Williams (Ed.) *Measuring the information society.* Newbury Park, CA: Sage.

Glaser, B. G., & Strauss, A. L. (1967). *The discovery of grounded theory.* New York: Aldine.

IC^2 Institute. (1989). *Consortia Database.* Austin, TX: IC^2 Institute, University of Texas.

Kincaid, D. L. (1979). *The convergence model of communication.* Unpublished paper, Honolulu, East-West Communication Institute.

Lawrence, P. R., & Lorsch, J. W. (1967). *Organization and environment.* Boston, MA: Harvard Graduate School of Business Administration.

Martin, P. Y., & Turner, B. A. (1986). Grounded theory and organizational research. *The Journal of Applied Behavioral Science, 22,* 141-157.

Peck, M. J. (1986). Joint R&D: The case of the Microelectronics and Computer Technology Corporation. *Research Policy, 15,* 219-231.

Petit, M. J. (1987). *Industrial research and development consortia.* Austin, TX: IC^2 Institute, University of Texas.

Rogers, E. M. (1983). *Diffusion of innovations* (3rd ed.). New York: Free Press.

Smilor, R. W., Gibson, D. V., & Avery, C. (1989, Spring). R&D consortia and technology transfer: Initial lessons from MCC. *Journal of Technology Transfer, 14*(2), 11-22.

Terreberry, S. (1986, March). The evolution of organization environments. *Administrative Science Quarterly,* 590-613.

Weick, K. E. (1988, August). *Interpretation systems as a context for new technologies.* Discussion draft prepared for Conference on Technology, Carnegie-Mellon University.

Williams, F., Avery, C. M., Poteet, B., Sawhney, H., Stewart, G., Gibson, D., Wey, J. Y. J., Nagasundaram, M., Brackenridge, E., & Dietrich, G. (1988). *Communication studies of technology transfer.* Working paper, Center for Research on Communication Technology and Society, University of Texas at Austin.

5

University and Industry Linkages: The Austin, Texas, Study

G. HUTCHINSON STEWART
DAVID V. GIBSON

Technology transfer between industry and universities involves the movement and application of knowledge. An examination of the structures through which universities and industry interact is essential when exploring the phenomenon of university-industry technology transfer. Before the actual transfer of technology can take place, linkages must be formed. Throughout the history of the United States, communication and cooperation between universities and industry has ebbed and flowed according to the needs of society. This chapter takes the position that, in the 1990s, as environmental uncertainties grow and the role of technology becomes more important to the economic well-being of the United States, the synergy of university-industry relations is being revitalized. Gary Stewart is a Ph.D. candidate in the Management Science and Information Systems Department in the Graduate School of Business at the University of Texas at Austin. He has more than 7 years of experience as a manager of international

support operations in a major U.S. high-technology firm. David
V. Gibson's institutional affiliations are listed in the introductory
chapter to this book.

Eras of the University

Throughout history, the world's universities have served as reposi-
tories for the accumulated works of researchers and scholars, and they
have played a central role in the processing and transfer of information
(Ackerman, Owens, & Pirkle, 1990). In the United States as elsewhere,
however, the perception of the value and the proper role of universities
has gone through cycles. At different points in a society's history there
are reasons why universities and industry have close associations and
why they do not. One explanation is that during times of uncertainty and
challenge there will be increased cooperation between a nation's universities and industry. This explanation fits with resource-dependency
theory, where the focal organization (the university) must attend to the
demands of those organizations or institutions in the environment that
provide necessary and important resources for the focal organization's
continued survival. Organizations will and should respond to the demands of those organizations that control critical resources (Pfeffer &
Salancik, 1978).

During the Civil War, President Lincoln signed into law the Morrill
Act, which created the mechanism for land-grant colleges (Thackery,
1971). Close cooperation between the land-grant institutions and the
public gave rise to a period of great technological advances in agriculture. In 1887, the Hatch Act provided funds for agriculture-experiment
stations, which were usually located at land-grant institutions. This act
promulgated university research in the agricultural sciences. Although
research at land-grant institutions continued and branched into scientific research, the culture of having universities and industry work
closely together decreased as the U.S. economy became less dependent
upon agriculture in the industrial age (Ashby, 1972).

Baba (1987) notes that, in the United States, cooperative university-
industry associations have formed in clusters of time centering around
periods of intensive international competition or crisis, and periods of
pronounced technological change. He describes four periods wherein
there has been an effort to coordinate the resources of the university and

industry: prior to and during World War I (1900-1916); during the Great
Depression (1929-1933); during and after World War II (1943-1954);
and during global competition (1967-present).

Harcleroad and Ostar (1987) identify four technological periods that
have occurred in the United States, which help explain the historical
periods of university and industry relations:

(1) the *agricultural society* (1776-1850) with land as the key resource,
localized transportation, and dispersed settlement patterns of population;

(2) the *manufacturing society* (1850-1920) with raw materials as the key
resource, transportation by water and rail, and the beginnings of population
centers in larger cities;

(3) the *service society* (1920-1960), with people as the key resource, transportation by automobile and airplane, and huge population centers giving
birth to the metropolis; and

(4) the *communications/international society* (1960-present) with information and knowledge as the key resources, and giving birth to the concept
of the technopolis and global information networks (Gibson, 1990;
Smilor, Kozmetsky, & Gibson, 1988).

Two late 20th-century trends are combining to encourage increased
cooperation between universities and industry. First, the United States
has been losing economic ground to its international competitors, primarily the Japanese (Abernathy & Rosenbloom, 1982). Second, technology is being viewed as a generator of wealth more powerful than
labor, resources, or capital (Gomes, 1989; Kozmetsky, 1988), and as
the central factor in the determination of world trade patterns (Terpstra,
1978). As a result of these trends, the role of the university in helping
to restore the nation's competitive position increasingly is being re-
examined (Gibson & Smilor, 1988; Hillenbrand, 1989; Sims, 1988;
Weinberg & Mazey, 1988).

University-Industry Relations in the 1990s

Pressures for reducing the national debt, balancing the federal budget, and lowering trade deficits have reduced the amount of federal
money available for university research (Elder, 1988). According to
Kennedy (1986) the cutbacks of federal fellowship and training-grant
programs to universities have meant that students are working on

outdated equipment, and that professors are having to devote increasing amounts of their time to soliciting money to finance operations and to equip their laboratories. In 1989, a faculty researcher had to receive 1.5 times the amount of research funds as in 1979 to support the same research results (Beyers, 1989). Such funding limitations have led many to view the professor not only as engineer or inventor, but as entrepreneur as well (Kouwenhoven, 1981). Faced with budget constraints, U.S. universities are searching for new sources to fund research and other academic programs (L. Jones, 1985).

An indication that the opportunity for commercialization of university research is receiving more emphasis is the fact that from 1965 to 1989, membership in the Society of University Patent Administration has grown from 60 to 500 members. Paralleling this trend is a decline in corporate R&D spending; for the first time in 14 years, 1989 saw U.S. spending on corporate research and development decline vis-à-vis inflation (Markoff, 1990). In their search for funding to maintain or enhance the quality of research and teaching programs, universities are becoming more dependent on industry funding. On the other hand, companies are looking to universities as reservoirs of basic and applied technology (Crudele, 1990) and new industrial ventures as well as established firms look at the quality and availability of university research as one of the major factors directing where a high-tech facility will locate (Gibson & Rogers, 1988).

As federal levels of research funding remain low or even decrease, and as U.S. technological competitiveness remains threatened, the perceived benefits of university-industry alliances will become all the more attractive to policymakers (Low, 1982). On the other hand, many who work in academia fear that as industry dedicates more of its resources to universities, there will be more of a desire on the part of industry to direct university operations: Academic freedom—a hallmark for U.S. universities—could be threatened, and science pressed into the service of capital (Noble, 1977).

Dating from the founding of the nation, there is historical precedence for the U.S. government to influence public policy through research and development activity (Link & Bauer, 1987). Close collaboration between the government, universities, and agriculture dates to the mid-1800s. In the industrial sector, however, U.S. universities traditionally have concentrated on basic research that has been commercialized, or applied, by industry at some future time. This time gap is considerably greater in the United States than in some foreign countries that compete quite well against the United States. In an effort to speed the commercialization of U.S. research, several laws have been enacted with the purpose of encouraging the formation of cooperative efforts among U.S. companies and between U.S. companies and universities (see Appendix A; Collins & Tillman, 1988). For example, prior to the passage of the Federal Technology Transfer Act in 1986, there was little incentive for university researchers to attempt to commercialize their technology (Ling & Wallace, 1989).

Linkages as the Unit of Study

Communication between industry and universities can be achieved through many methods. Auster (1989) identifies three units of analysis for the purpose of studying interorganizational relations: linkages, organization sets, and networks. Linkage activity requires that people cross organizational and institutional boundaries to interact with personnel in other organizations and institutions. As noted, environmental uncertainty is one critical dimension that drive the formation of interorganizational linkages (Auster, 1990; Katz and Kahn, 1966; Pfeffer & Salancik, 1978). Cutler (1988) identifies people linkages (along with publications and patents) as one of the three important elements of technology transfer. Kwiram (1989) states that the "linkage laboratory" is the missing component between the university and industry in the technology transfer process.

While research into linkages and their impact is still in its infancy, Stewart (1990) identifies 72 linkages by which professionals from industry and academia interface (Appendix B). A linkage may or may not be used depending upon the predisposition of either industry or academic professionals. In order to determine whether a particular linkage is effective (a facilitator) or not effective (an inhibitor) in the communication process between the university and industry, it is useful to examine the attitudes of industry and academic personnel and the cultures of their respective organizations.

Findings from the Case Study of Austin, Texas

Austin, Texas, is a developing technopolis (Smilor, Gibson, & Kozmetsky, 1988), having many major U.S. electronics companies and their R&D facilities located in the area along with the research consortia MCC and Sematech and the University of Texas at Austin. The present research is based on interviews with employees from the Austin divisions of three technology-based firms, and faculty from the University of Texas at Austin.[1]

Classroom Linkages

The need for professionals in high-technology industries to have access to the latest research and technological capability for their specialties is real and pressing. Advances in technology are occurring so rapidly that individuals graduating with the most up-to-date technical educations will find that the amount of knowledge in their fields literally turns over within a few years. The half-life of education in today's electronic technology is three to four years. People can become obsolete technologically very quickly if they do not take refresher courses in their given specialties. As a result, many research and technology companies expect their employees to update their educational status on an ongoing basis. Of the companies studied in the present research, two require their employees to have 40 hours of real-time class instruction per year toward their professional development. One company is now asking all of its technical personnel to get at least a bachelor's-level degree. In addition to the demands of employers, many employees seek incremental education for personal goal fulfillment. Industry professionals commonly expect major local universities to provide access to evening degree programs in advanced classes in their fields of expertise.

Reinforcing the value of face-to-face instruction, members of industry often view the classroom experience as a valuable networking tool where personal, informal associations are formed among industry professionals and between university and industry. These linkages often lead to future benefits, such as employment opportunities and technology transfer; as one professional said, "a lot goes on in class besides teaching." Linkages formed during classtime hours may well pay future dividends for both the university and industry. In the course of the present research we learned of several important, long-term U.S. university and industry linkages that were initiated in the classroom.

With the exception of laboratory work, many industry professionals believe that students can learn just as well from video as from in-class instruction. This belief (if substantiated) can have profound implications for universities. Those universities that are in the business of videotaping and making accessible the lectures of their best professors will have a large and appreciative market for their services. Those universities that neither offer evening or weekend courses nor sell videotape services could suffer a potential loss of industry goodwill and support.

Research Linkages

Industry often minimizes the teaching function that is performed when students at universities participate in the research process. In industry, research is commonly viewed as a means to some application, and industry generally looks to universities for research that ultimately will lead to financial benefit. In academia, research often is viewed as an end in itself. Academics feel strongly that a very important part of the teaching process includes teaching good research techniques, so that "research for research's sake"—often disdained by industry—has a valid and necessary place in the university setting.

Three common views of universities are held by industry, and universities are ranked from excellent to poor in each category. First, there is the *teaching institution,* where industry is assured that students are getting excellent instruction and graduates will make good employees. Second, there is the *application-based institution,* where students receive hands-on training with sophisticated technologies. Third is the *theoretical research institution,* which is viewed as producing undergraduates and graduates that may not be ready for hands-on work when they graduate, but are well schooled in the theoretical aspects of their subject matter. Some universities are excellent in all three categories, and some concentrate their resources on selected aspects of the university's mission (Stewart, 1990).

Depending upon the particular needs of a company, the reason for wanting an association with a university will vary. In some instances, the need for research results in the motive for association. As one industry professional stated, however, his company received less than 1% of its basic research from its ties with universities. His company seeks strong university ties so that it will have access to good undergraduate and graduate students. Similarly, many companies equip university labs so that students, upon their graduation, will be familiar with the equipment and can "hit the ground running" upon their hire by the company.

One method by which universities enhance their national and international reputations is the pursuit of excellence in "key" or "super" technologies. Universities pursue technology excellence because the financial and prestige payback can be substantial. University administrators are increasingly aware of the importance of "top 20 lists" in critical technical areas, as these lists are one potential culling device for determining which universities are perceived to be conducting the most exciting research and attracting the best professors and students. For example, the University of Houston is known as a world leader in superconductivity, and Dr. Paul Chu has received over $30 million in funding from various federal, state, and local sources to achieve a fourfold increase in superconductor densities (Stanley, 1989). Early technological leadership is important, because any one key technology probably cannot support more than four or five research centers. It is impractical for 20 to 30 universities to have visions of becoming centers of excellence for a given technology (Stewart, 1990).

While pursuing excellence in research (often thought of as basic research), universities often shy away from establishing hands-on applied-type relationships with industry. Several industry contacts stated that there is no reason why universities cannot be both excellent basic research centers and have the capability to meet the other needs of industry. Often-cited examples of universities that meet both the basic and applied needs of industry are Stanford University and MIT. Furthermore, student researchers do not have to be at the graduate level before they can make important contributions to research and to industry. One group of undergraduate students at the University of Texas at Austin, taking a required research course in mechanical engineering, invented and patented a device to "scrub the air" and remove dangerous components of acid rain from industrial smokestack emissions. Thus, teaching and research combined in a positive way for faculty, students, and industry, producing a win-win situation for all involved.

Publication Linkages

The university mandate to publish research results ("publish or perish") and the private sector's desire to maintain company secrets for competitive advantage are important and often contradictory issues that can form a barrier to university-industry technology transfer. Collaboration between the university and a company can be delayed, or voided, due to this barrier. Professional advancement for academics depends upon the quality and frequency of publication in prestigious journals. Companies, on the other hand, often want to delay as long as possible the public dissemination of knowledge.

The present research noted two university-industry institutional activities that encouraged the early publication of research results. Management in one of the firms studied has come to view the publication of articles describing the application of its products as an important marketing and public relations tool. The belief is that the more that articles about its products appear in prestigious publications, and the earlier they appear, the better the company's sales of those products will be.

In a second example of close industry-university cooperation, one company found a way to enhance its penetration into university laboratories through a collaborative effort involving the publication of a state of the art technology-based textbook. Normally, the lag time between the introduction of a new technology and the publication of a textbook detailing academic constructs of the product is from two to three years at a minimum. By involving a University of Texas professor in the early stages of a semiconductor chip design, the company was able to get the technology-based textbook in the classroom in only three months after introduction of the product into the marketplace.

Financial Linkages

There are three basic financial linkages through which companies interface with universities: gifts, grants, and contracts. A financial gift enables the university to use the funds for whatever purpose it desires, without consulting the company; universities especially like to receive financial gifts. A grant is a transfer of funds to be used at the discretion of the university, but often within a certain subject area or academic department (for example, within a college of engineering, but for any project that the college desires). A contract is an agreement by a company to provide funding for a specific project or purpose. A report detailing the use of grant or contract money commonly is provided by the university to the company.

Grants and contracts have proven to be a source of friction between universities and companies. According to the institutional norms of universities, a certain percentage of a grant or contract must be dedicated to the general overhead charges of running the university, and thus not be applied directly to the project that the company is funding. Overhead charges of over 40% are not uncommon (Stanford

University's indirect cost rate is 74%), and some members of industry and the academia believe that such percentages are too high. The subject of indirect costs to universities is an increasingly controversial issue for the federal government, industry, and academic institutions: "They (government and industry) think of it as a procurement rather than an investment" (Horton, quoted in Shurkin, 1990, p. 20). And the issue is causing strain between the sciences and humanities as to who is subsidizing whom: "To pit one part of the faculty against the other is shortsighted, divisive, and unreasonable. The problem won't get fixed without cooperation between the faculty and administration. This sense of collegiality and unity is doubly important as we face a new world" (Kennedy, quoted in Shurkin, 1990, p. 20).

While university overhead charges are of concern to many U.S. companies, overhead charges of 40% are not uncommon in industry-to-industry ties. Furthermore, as institutions of higher learning, universities have broad-based needs. There are many and varied academic disciplines and programs that are essential to the culture and quality of a university education and which traditionally have not benefited from broad-based industry support. U.S. universities would not be the valuable resources they are today if these fundamentally important, but not profit-oriented activities were left with insufficient funds. It is also interesting to note that overhead charges have not been an impediment to many Japanese companies executing research contracts with U.S. universities (Stewart, 1990).

Both professionals in industry and university professors indicate that the bureaucracy involved in university-industry communication and technology transfer processes often results in frustratingly slow action (M. Jones, 1985). While industry is often anxious to have the university-research relationship progress at a rapid pace, a university is faced with many legal and institutional constraints that need to be addressed, especially during times of shifting norms and values which impact the entire university (Brett, Gibson, & Smilor, 1990). Once again, many Japanese companies are not adverse to taking a long-term approach to instituting ongoing associations with U.S. universities (Stewart, 1990).

The Linkage Champion and
International Technology Transfer

While many in industry and academia agreed that industry-university ties are very important, it is unusual to have a full-time corporate-based person budgeted for the function. The job of interfacing the company with universities is often an additional task given to an already busy executive. The job is done on a part-time basis if it is done at all. On the other hand, research has shown that it takes a dedicated "linkage champion" to accomplish the movement and application of knowledge from one entity to another (Smilor & Gibson, 1990). Some person has to be dedicated to such transfer, and has to be a champion of the process in order for it to be successful. Success is not likely to occur if the university-industry interface is "another" duty piled on many others.

In the United States, when a person goes from industry to teach or engage in research at a university there is often no professionally and personally rewarding way for that individual to return to industry. The same situation often applies to university professors who take leaves from academia to work in industry. Industry and university promotion and reward systems often inhibit personnel from being boundary spanners even if such interchanges would be beneficial for the companies and for the universities (Stewart, 1990). As one professional put it, going to work at a university "is the kiss of death for a career in industry." Consequently, in the United States, when people move from industry to the university, it usually is at the end of their careers.

According to Cutler (1988), Japanese—in both industry and university settings—are more likely to cross organizational boundaries to gain information than are their U.S. counterparts. For researchers attending technical meetings outside their work location (at least twice each month), Cutler found that Japanese university personnel were twice as likely as their U.S. counterparts to cross organizational boundaries and Japanese industry personnel were found to be more than four times as likely than their U.S. counterparts to cross organizational boundaries (Table 5.1).

U.S. firms increasingly are becoming concerned that foreign companies are making too many important linkages with U.S. universities through financial contributions and foreign students. Of major concern are exclusivity agreements, wherein research results are given exclusively to the foreign companies funding the research. Not only are U.S. companies denied access to these research results, but the U.S. professors' research talent may be tied up with work for foreign competitors.

Some U.S. industry professionals, as well as some academics, are concerned about the number of foreign students in the technical disciplines in U.S. universities. Others point out that there are not enough

TABLE 5.1: Percentage of U.S. and Japanese University and Industry Personnel That Cross Organizational Boundaries

	United States	Japan
University Personnel	43	93
Industry Personnel	17	80

SOURCE: Cutter, 1988

qualified U.S. applicants to fill all the available openings in U.S. graduate schools, and that U.S. universities benefit by accepting the most highly qualified students regardless of what countries they come from. It is also argued that U.S. industry would also benefit if these foreign students were encouraged to work in U.S. firms (whether at home or abroad).

A contributing factor to the high numbers of foreign graduate students in U.S. universities is the fact that foreign companies often pay these student-employees while they pursue advanced educations in the United States. After they complete their educations, these student/employees are encouraged—if not required—to return to the subsidizing companies and use their U.S.-based education and network links. U.S. companies typically do not pay employees while they pursue advanced education on a full-time basis, and several respondents gave examples where U.S. companies did not utilize employees' advanced degrees once they were obtained.

Conclusion

In keeping with a resource-dependence view, as financial resources become limited universities will seek to establish closer ties with industry, and as quality academic resources become more valued private companies will become more focused in the way that they distribute funds for university research and education. Another important motivation for increased university-industry cooperation is the increasingly global and competitive race to develop and commercialize new technologies. Together these environmental shifts have important implications for both U.S. industry and universities.

Universities that do not form cooperative alliances with industry may find themselves restricted in their search for major sources of funds for research and teaching. Industry would benefit from a commitment to form cooperative linkages with universities to facilitate technology transfer through education and research. On the other hand, there is the concern that U.S. industry's pursuit of short-term profit not be allowed to weaken the institutional norms and values that have facilitated the development of the existing excellence of U.S. universities, a resource which many countries envy.

U.S. universities can benefit from a strong and competitive U.S. industrial sector that supports quality education and research, and excellent universities can contribute to the competitive strength of U.S. industry. Through its short history, the United States has witnessed the benefits of such synergy. Most recently in mature and emerging technopoleis (such as Silicon Valley, California; Route 128, Massachusetts; Research Triangle, North Carolina; and Austin, Texas) there are vivid examples of where such cooperation has enhanced both university excellence and the global competitiveness of U.S. industry. The present research emphasizes the value of the effective use of institutional and personal linkages to build cooperative alliances between U.S. universities and industry as a necessary and important step to improving the nation's intellectual and technological competitiveness in the global economy.

$$\boxed{\text{A}}$$

Federal Legislation Facilitating
University-Industry Cooperation

1954—Section 174 of the Internal Revenue Service Code allowed full deduction of R&D expenditures. (Link & Bauer, 1987.)

1980—The U.S. Department of Justice issues the opinion that "the closer joint activity is to the basic end of the research spectrum . . . the more likely it is to be acceptable under the antitrust law." (U.S. Department of Justice, 1980.)

1980—Stevenson-Wydler Innovation Act of 1980, Public Law 96-480. Elevated status and importance of technology transfer by mandating an office of Research and Technology Application at all major federal laboratories. (Gillespie, 1988.)

1980—Bayh-Dole Act of 1980, Public Law 96-517. Permitted nonprofit organizations and small businesses that receive federal funding the option of keeping title rights to inventions they develop. (Gillespie, 1988.)

1980—Patent and Trademark Amendments of 1980, Public Law 96-517. Gave universities rights to federally funded inventions developed at the universities. (Erickson & Baldwin, 1988.)

1982—Small Business Innovation Development Act, Public Law 97-219. Requires federal agencies with research grant or contract programs exceeding $100 million to "set aside" 10% of the funds for R&D by small businesses. (Waugaman & Martin, 1985.)

1983—President Reagan issued a directive to federal agencies to give the advantages of the Bayh-Dole Act of 1980 to larger companies as well. (Gillespie, 1988.)

1983—The White House Science Council's Federal Laboratory Review Panel recommended that each federal laboratory have an external oversight committee with strong industry and university representation. (Ling & Wallace, 1989.)

1984—Amendment to Patent and Trademark Amendments Act of 1980, Public Law 98-620. Extended coverage, and reduced restrictions of the 1980 Act. (Erickson & Baldwin, 1988.)

1984—National Cooperative Research Act of 1984, Public Law 98-462. Established a rule of reason for evaluating the antitrust implication of each R&D joint venture, and limits potential liability to actual, not treble damages. (Link & Bauer, 1987.)

1984—Joint Research and Development Act of 1984. Encourages joint ventures between private companies. (Link & Bauer, 1987.)

1986—Federal Technology Transfer Act of 1986, Public Law 99-502. Permits government-owned and operated laboratories to enter into joint R&D agreements. (Gillespie, 1988.)

1986—Tax Reform Act of 1986. Extended tax credits for R&D through December, 1990. (Erickson & Baldwin, 1988.)

1987—President Reagan directed agencies responsible for government-operated laboratories to delegate authority to those laboratories for entering into R&D agreements, and licensing of intellectual property. (Gillespie, 1988.)

1987—Omnibus Trade and Competitiveness Act of 1987. Attempts to improve manufacturing and product technology for United States companies. (Collins & Tillman, 1988.)

1988—Superconductivity Competitiveness Act of 1988. Allows collaborative government-industry R&D results trade-secret protection for up to 5 years. (Gillespie, 1988.)

B

Examples of University-Industry (UI) Linkages

Adjunct Professional—Similar to the adjunct professor, except that in this case a full-time faculty member of the university is working on an adjunct basis in industry.

Adjunct Professors—Close in concept to consulting, wherein a professional from industry is also a part-time member of the faculty at a university, while still maintaining his or her professional position in industry. (*Executive loan* is a concept very similar to the adjunct professor).

Advisory Board—Members of industry sit on panels in order to share their professional expertise with their counterparts in academia.

Alpha and Beta Site Linkages—A university would receive prototype parts from a company, and perform evaluation, and give feedback to the company, as the company continues with its product development.

Alumni Associations—Alumni from a university and who are personnel of a company within industry maintain ties with their university.

Alumni Loyalty—Wherein a member of industry who is an alumnus of a particular university would "think first" of his or her alma mater when the need for a linkage with an academic institution arises.

Annual Reports—Publications from industry, published annually, which can create interest on the part of, and contact from, university personnel.

Applied Research Linkages—Cooperation between industry and the university for the purpose of establishing an applied research effort.

Associations—Professional associations, wherein professionals from both industry and academia may be members, and share information during association meetings.

Athletic Events—Members of industry and academia may come together and further their associations while attending these events.

Basic Research Linkages—Cooperation between industry and the university for the purpose of establishing a basic research effort; can be general (open) or directed (targeted).

Breakfasts—Regularly-scheduled meetings (as opposed to the one-time conference or seminar) wherein information exchanges can take place between personnel from the university and industry.

Classroom Interface—Wherein students who are members of industry can talk before, during, and after class with their professors concerning projects that may be beneficial to both.

Clubs—Organizations of students specializing in a particular academic area (for example, a "real estate club") who might have contacts with a similarly interested company or companies in the private sector.

Common Alma Mater Alumni—Members of industry working for the same company, who were not actively connected with their former university, but who, as a result of the synergy from two (or more) of them, form a linkage with the university.

Conferences—Information is presented by either industry or the university, for the other party; e.g., university-sponsored conferences (or seminars) or industry-sponsored conferences (or seminars).

Consulting—Usually of the type wherein university personnel are consultants for industry.

Correspondence Courses—By-mail courses, where students working in industry take courses from the university.

Curriculum Development—Individuals from industry work with members of academia to help develop curriculum content.

Custom Courses—A university, at the request of a company, sets up a course to meet a specific and special need of the company.

Data Bases—Information indices of innumerable types which can be accessed by both industrial and academic personnel, and which can be sourced from either academia or industry.

Designated Liaison—A company appoints a person (usually, but not necessarily, an alumnus) to be the focal point for communication between the company and a particular university.

Education Programs—Employees from industry are taking courses provided by the university (e.g., regular university course offerings; special degree programs; university instructors at the industry site; videotape or satellite instruction).

Employee Rotation—An employee may rotate employment from industry to academia and back to industry, and may in fact rotate to different locations within both industry and academia. Differs from executive loan in that the loan is a one-time discrete event, and rotation may be continuous.

Equipment Donation—A company gives equipment to the university for use in university operations. Differs from hardware parts, in that these donations are usually larger and involve equipment worth more money.

Equipment Expertise—A member of industry, or a member of a university, due to his or her particular knowledge about a certain or piece of hardware, is part of a linkage to help the other entity learn to operate the equipment or keep the equipment running.

Executive Loan—An individual from industry goes to a university (is, in effect, "loaned") for a certain period of time. Differs from adjunct professor in that the executive is "full time" at the university during the loan period, and does not perform a function in both industry and the university at the same time.

Employment of Current Students—Employment of students either in summer programs, or on a part-time basis, as students pursue their education.

Former Students Network—Knowledge of a university and its capabilities gained by a student when he or she attends the university, and which later may be used by the former student when working in industry.

Founders—The men and women who establish companies, and who often have a loyalty to the academic institutions where they obtained their degrees.

Full-Time Employed Students—A student who is on the payroll of an industrial company goes to school full-time and gets a degree, and then goes to work for the company after completing the degree.

Funding from Industry to the University—Grants, research money, or other programs in which funds flow from industry to the university.

Governmental Appointments—Appointments of industry and academic personnel to committees that have as their function an interface between the academic and industrial communities.

Guest Speakers—Members of industry visit a university class, or club, or some organization within the university to give a presentation on the topic of their expertise.

Hardware or Software Sample Programs—Programs wherein industry provides the university with physical parts, or software, to aid the university in teaching or research.

Hiring University Graduates—Industrial concerns hire graduates of a particular university, thus establishing or furthering a linkage.

Indirect Research Linkages—A company (perhaps a foreign concern) has a research association with a university, and a second corporate entity might gain access to technology of the first company, through its own ties to the university.

Internships—Students work for a company on a continuing but not full-time basis such as in the summers, or during part of a day, usually while continuing their education, and then go to work full-time for the company after completing their education.

Intensive Short Courses—Course offerings which are reduced in length (though not in content—perhaps a week or less) and given to industry employees by the university, to allow employees to get the course without having to miss as much time from work.

Intradepartmental Ex-Students—Former students who have a strong bond to, and maintain contact with, a particular department at a university.

Journal Exchanges—Discussions between members of industry and the university for the purpose of sharing information concerning articles written in academic or industry journals and other publications.

Key Programs—Programs at various universities—for example, superconductivity and X ray lithography—which may be a specialty of a university, and which may attract the interest of industry.

Library Access—The libraries of a company and those of the university have agreements to share and exchange information.

Luncheons or Dinners—Similar in structure, of course, to breakfasts, but there may be a speaker, and more formality to the occasion.

New Course Development—Industry works with the university to create a totally new course that has not ever been offered because the course teaches new technology. Differs from curriculum development in that it is not selecting a course, but creating one.

Papers—Presentation of academic papers at conferences, conventions or other meetings by academics or professionals from industry, where there can be exchange of ideas.

Patents—Joint university-industry efforts for securing patents.

Product Contests—Companies solicit entries for original and unique uses of their products, and these contests are open to university students and professors, in addition to professionals in industry.

Professorial Summer Study—A member of academia "trains" in an industrial environment during the summer months between semesters of school.

Project Monitoring Groups—Joint UI groups that follow the progress of any cooperative ventures between industry and the university.

Recruiting Visits—Visits to academic institutions made by members of industry that may result in beneficial linkages.

Religious Institutions—Members of industry and academia share a common religious meeting place, or belief, and a linkage is formed.

Research Council—Councils consisting of members from both academia and industry that meet periodically for the specific purpose of discussing and selecting important research topics.

Research Papers—Those papers appearing in academic journals or trade publications that lead to a linkage between academia and industry.

Retirees—Persons who after retirement still maintain ties with industry, and perhaps with the university, and can be part of a linkage between the university and industry.

Sampling—Industry sends new and functional parts on a regular basis to the university for its use. Differs from alpha site linkage in that sampling programs utilize commercial-quality parts, not parts still being developed.

Satellite Courses—Students from industry who are taking a course via satellite form a linkage with the professor or university that produced the course.

School Rankings—The top-rated colleges and universities in a given discipline (for example, the top 20 engineering schools) would receive more attention from industry as a result of being on the list of highly ranked schools.

Semester Projects—Work done for industry by either students under professorial guidance, or by professors themselves, which may or may not have a term paper as the result.

Seminars—Short-duration presentations, usually of one or two days, given by universities, and which may include industry personnel on the invitee lists.

Social Systems—University fraternity and sorority systems, where long-lasting relationships that carry on into industry may be developed, and may be related strongly to the university through alumni organizations.

Special Events—On-campus occasions, such as "Parents' Day," wherein members of industry may play a part in the proceedings.

Special Services—Services provided by the university for industry, such as specialty lab work, and that result in or furthers a linkage.

State-Encouraged Linkages—Mandates by state legislatures that universities and industry cooperate in order to achieve the movement of knowledge from one entity to the other.

Student Internships—Students are employed during their educational process, much as doctors intern in hospitals. Also known as "student co-op programs."

Symposium—An industrial company holds an affair to acquaint professors with its products, with the goal of gaining increased awareness and usage among those academics, and within their respective institutions.

Term Papers By Students—Students doing "course projects" for a company (usually arranged between the students' professor and industry) write up the results of the semester in a formal report.

Textbook Authorship—Associations wherein a professor writes a text for a company in private industry, or co-writes a textbook with an industry person.

Trade Shows—Industry events, where papers may be presented by either industry or academic personnel, and wherein linkages may be started.

University-Industry Equipment Access—Industry personnel are allowed access to university equipment or facilities, or university personnel are allowed access to industry equipment or facilities.

University Relations Departments—Dedicated departments within organizations that are utilized to further relations and linkages between the company and universities.

Visiting Committees—Industry professionals from (perhaps different) companies serving on advisory boards for academic institutions, to help formulate policies in areas such as curriculum development.

Note

1. Data for this ongoing research are based, in part, on interviews with 34 industry and academic professionals. Data for the industry component were collected on three high tech firms with major production and R&D operations located in Austin: Texas Instruments, Motorola, and 3M. Professionals from all three companies made up the industry interview base. Texas Instruments is a Dallas-based company with a major facility located in Austin since 1969. Motorola is a Schaumburg, Illinois-based company, with a semiconductor operation in Austin since 1974. 3M, headquartered in Minnesota, began moving several R&D divisions to Austin in 1984. Interviews were conducted with University of Texas at Austin professors from the electrical and computer engineering, mechanical engineering, and computer science departments.

References

Abernathy, W. J., & Rosenbloom, R. S. (1982). The institutional climate for innovation in industry: The role of management attitudes and practices. In A. H. Teich & R. Thornton (Eds.), *Science technology, and the issues of the eighties: Policy outlook.* Boulder, CO: Westview Press.

Ackerman, D., Owens, R., & Pirkle, K. (1990). *The role of Old Dominion University in economic development.* Unpublished manuscript, Old Dominion University, Norfolk, VA.

Auster, E. R. (1989). *An exploratory analysis of the relationship of industry uncertainty to birth rates of interorganizational forms: Corporate linkages between U.S. and Japan 1984, 1985.* Unpublished manuscript, Columbia University, New York.

Auster, E. R. (1990). *Bringing a multilevel perspective into research on interorganizational relationships.* Unpublished manuscript, Columbia University, New York.

Ashby, E. (1972). *The campus and the city.* Berkeley, CA: The Carnegie Foundation for the Advancement of Teaching.

Baba, M. L. (1987). University innovation to promote economic growth and university/industry relations. In P. A. Abetti, C. W. LeMaistre, R. W. Smilor, & W. A. Wallace (Eds.), *Technological innovation and economic growth: The role of industry, small business entrepreneurship, venture capital, and universities.* Austin: IC^2 Institute, University of Texas at Austin.

Beyers, B. (1989, January). For love of science. *Stanford Observer,* p. 1.

Brett, A., Gibson, D., & Smilor, R. (Eds.). (1990). *University spin-off companies: Economic development, faculty entrepreneurs, and technology transfer.* Totowa, NJ: Rowman & Littlefield.

Collins, T. C., & Tillman, S. A., IV. (1988). Global technology diffusion and the American research university. In J. T. Kenney (Ed.), *Research administration and technology transfer.* San Francisco, CA: Jossey-Bass.

Crudele, J. (1990, February 5). U.S. can't afford to lose technology war. *Austin American-Statesman*, p. CB14.

Cutler, R. (1988). Survey of high-technology transfer mechanisms in Japan and the USA. *The Journal of Technology Transfer, 13*(1), 42-48.

Elder, M. (1988). The future of university research administration. In J. T. Kenney (Ed.), *Research administration and technology transfer,* San Francisco, CA: Jossey-Bass.

Erickson, G. A., & Baldwin, D. R. (1988). The new frontier of technology transfer. In J. T. Kenney (Ed.), *Research administration and technology transfer.* San Francisco, CA: Jossey-Bass.

Gibson, D. (Ed.). (1990). *Technology companies and global markets: Policies and strategies to accelerate entrepreneurship and innovation.* New Jersey: Rowman and Littlefield.

Gibson, D., & Rogers, E. M. (1988). The MCC comes to Texas. In F. Williams (Ed.), *Measuring the information society.* Newbury Park, CA: Sage.

Gibson, D., & Smilor, R. (1988). The role of the research university in creating the technopolis. In A. Brett, D. Gibson, & R. Smilor (Eds.), *University spin-out companies: Economic development, faculty entrepreneurs, and technology transfer.* Totowa, NJ: Rowman & Littlefield.

Gillespie, G. (1988). Federal laboratories: Economic development and intellectual property constraints. *The Journal of Technology Transfer, 13*(1), 20-26.

Gomes, S. (1989). Revitalizing America's competitiveness in the world. In F. Borkowski & S. McManus (Eds.), *Visions for the future.* Tampa: University of South Florida Press.

Harcleroad, F., & Ostar, A. (1987). *Colleges and universities for change.* Washington, DC: AASCU Press.

Hillenbrand, M. (1989). Policy issues in a global society. In F. Borkowski & S. McManus (Eds.), *Visions for the future.* Tampa: University of South Florida Press.

Jones, L. (1985). *University budgeting for critical mass and competition.* New York: Praeger.

Jones, M. (1985, May 2). UT researchers entangled in state red tape on purchases. *Austin American-Statesman,* p. A16.

Katz, D., & Kahn, R. L. (1966). *The social psychology of organizations.* New York: Wiley.

Kennedy, D. (1986). Basic research in the universities: How much utility? In R. Landau & N. Rosenberg (Eds.), *The positive sum strategy: Harnessing technology for economic growth.* Washington, DC: National Academy Press.

Kouwenhoven, J. (1981). James Buchanan Eads: The engineer as entrepreneur. In C. Pursell, Jr., (Ed.), *Technology in America.* Cambridge: MIT Press.

Kozmetsky, G. (1990). The challenge of technology innovation. The new globally competitive era. In D. Gibson (Ed.), *Technology companies and global markets: Programs, policies, and strategies to accelerate innovation and entrepreneurship.* Cambridge, MA: Ballinger.

Kwiram, A. L. (1989). Innovative industry-university linkages: Is there a missing link in our economic development chain? In K. Walters (Ed.), *Entrepreneurial management.* Cambridge, MA: Ballinger.

Ling, J., & Wallace, G. (1989, Fall). Government-to-industry technology transfer: Is it working? *The Bridge,* 21-25.

Link, A. N., & Bauer, L. L. (1987). An economic analysis of cooperative research. *Technovation, 6,* 247-260.

Low, G. (1982, December). *The organization of industrial relationships in universities.* Speech, National Conference On University-Corporate Relations in Science and Technology, University of Pennsylvania, Philadelphia.

Markoff, J. (1990, January 23). A corporate lag in research funds is causing worry. *New York Times,* p. 1.

Noble, D. (1977). *America by design: Science, technology, and the rise of corporate capitalism.* New York: Knopf.

Pfeffer, J., & Salancik, G. (1978). *The external control of organizations.* New York: Harper & Row.

Sims, R. R. (1988). Higher education in corporate readaptation. In J. T. Kenney (Ed.), *Research administration and technology transfer.* San Francisco, CA: Jossey-Bass.

Shurkin, J. (1990, January-February). "$22-million budget cut launched." *Stanford Observer,* pp. 1-20.

Smilor, R., Kozmetsky, G., & Gibson, D. (Eds.). (1988). *Creating the technopolis: Linking technology commercialization and economic development.* Cambridge, MA: Ballinger.

Smilor, R., Gibson, D., & Kozmetsky, G. (1988). Creating and sustaining the technopolis: High technology development in Austin, Texas. *Journal of Business Venturing, 4,* 49-67.

Smilor, R., & Gibson, D. (in press). Accelerating technology transfer in R&D consortia. *Research Technology Management.*

Stanley, D. (1989, February 3). Texans nearing commercial use of superconductivity. *Austin American-Statesman,* p. B4.

Stewart, G. H. (1990). Large corporations and the research university: An examination of factors of technology transfer. Doctoral dissertation, University of Texas at Austin.

Terpstra, V. (1978). *The cultural environment of international business.* Cincinnati, OH: South-Western.

Thackery, R. (1971). *The Future of the State University.* Urbana: University of Illinois Press.

U.S. Department of Justice. (1980). *Antitrust guidelines concerning research joint ventures.* Washington, DC: Government Printing Office.

Waugaman, P. G., & Martin, E. (1985, Summer). Small business innovative research and academia: Is the menace an opportunity? *SRA Journal,* 13-23.

Weinberg, M., & Mazey, M. E. (1988). Government-university-industry partnerships in technology development: A case study. *Technovation, 7,* 131-142.

6

University and Microelectronics Industry: The Phoenix, Arizona, Study

ROLF T. WIGAND

A major context for technology transfer is between universities and industry, as described in this study of technology transfer within the microelectronics industry in the Phoenix, Arizona, metropolitan area. Dr. Wigand first describes the general setting of high-technology development in Phoenix, the roles of industry and university—Arizona State University (ASU)—as well as the research and development environment. He then reports selected results of a specific study that he and colleagues undertook to identify interorganizational cooperative efforts in microelectronics technology transfer between industry and the university. Dr. Wigand is director of the Program in Information Management, School of Public Affairs, Arizona State University.

AUTHOR'S NOTE: Portions of this research were supported by a grant (No. ISI 8505583) from the National Science Foundation. Any opinions, findings, conclusions, or recommendations expressed are those of the author and do not necessarily reflect the views of the National Science Foundation. The author appreciates suggestions and contributions received from Everett M. Rogers, University of Southern California, Los Angeles, CA; and Judith K. Larsen, Dataquest, San Jose, CA.

Transfer Among Organizations

Technology has had an increasing impact on the activities and relationships of industry, government, and universities, as well as individuals. Study after study indicates how the effective transfer of technology now is codetermined more often by these public and private actors. It is almost a cliché to state that the high-technology industry is one of today's newest high-growth industries. Bloch (1986) and numerous others have reported how many cities, states, and countries around the world seek out high-technology companies to locate within their respective jurisdictions. It is well documented (e.g., Larsen & Wigand, 1987; Wigand, 1985, 1988; Wigand & Frankwick, 1989) that one essential feature high-technology companies look for when evaluating potential new and relocation sites is the degree to which industry, universities, and government cooperate in research and technology development programs and thus, in turn, make some aspects of technology transfer possible and effective.

High-technology companies—and, in particular, microelectronics firms—desire cooperative research environments because they depend increasingly upon research and innovation to remain competitive in their markets. Cooperative research efforts include the exchange of information among the participants, making research findings more widely available, taking advantage of synergistic effects, and increasing the speed to reach end-user markets. Prominent examples of such cooperative research efforts to enhance technology transfer in the electronics industry are MCC and SEMATECH. Larsen and Wigand (1987) and Larsen, Wigand, and Rogers (1987), as well as Snyder and Blevins (1986), argue that industry and university research cooperation is important as the speed with which research ideas progress to market-ready products is increasing in significance to high-technology firms (because the time between product introductions and obsolescence is decreasing).

The technology transfer of the average product in the United States takes about seven years, although in Japan this typically takes only five years. Most experts tend to agree that this difference is not explainable by Japanese advances and differences in technology as well as know-how, but can only be explained by people—that is, the individuals involved in the technology transfer process. Recognizing this situation, high-technology industries are attempting to increase the speed of technology transfer and the amount of joint cooperative research with universities (e.g., Fowler, 1984; Snyder & Blevins, 1986).

Cooperative industry-university research can be viewed as a two-way information exchange process in which technological information resulting from research flows from the university to the industry, and from industry to the university. In this process, supportive government organizations and programs very often may encourage and facilitate the information flow between industry and the universities.

High-Technology Development in the Phoenix Metropolitan Area

Phoenix is the nation's 10th largest city, with 924,000 inhabitants in 1988; the metropolitan Phoenix area (Maricopa County) comprises more than two million people (Bureau of the Census, 1989), almost 70% of Arizona's population. High-technology manufacturing (computers, electronic components, aerospace, communications, and scientific instrumentation) amounts to almost 50% of all manufacturing jobs in Arizona. The comparative national average is 15%. Arizona has well over 300 high-technology firms, employing more than 85,000 workers. Metropolitan Phoenix high-technology firms employ more than 80% of the Arizona high-technology work force. High-technology growth in Arizona has averaged about 6%, compared to a national average of 2.4% annually. The largest high-technology firm, Motorola, employs more than 22,000 individuals. Although growth patterns have dampened at the time of this writing, one should visualize that during the last 17 years Arizona has added more jobs to its economy than France, the Federal Republic of Germany, Great Britain, Italy, and Sweden combined. Arizona was the fastest-growing state in the nation from 1975 to 1985 in terms of employment and personal income. In terms of population growth, Arizona ranked third nationally behind Alaska and Nevada.

High-technology industries, universities, and government work reasonably well together in Arizona. A report, *Arizona Horizons: A Strategy for Future Economic Growth* (Arizona Office for Economic Planning and Development, 1983), was issued by then-Governor Bruce Babbitt making specific recommendations pertaining to capital availability, entrepreneurial and small business support, education, and manpower training as well as research and development. This latter area addressed an increase in research funding and support, setting up a council on science and technology, developing a long-term state telecommunications policy, reviewing the Arizona Board of Regents' patent policy and sponsored-research policies, as well as exploring formal mechanisms to promote the transfer of technology between the state universities and businesses in the state and among the businesses themselves. Many of these efforts have been instituted since this 1983 report.

As high technology develops in the state, state government can play an important role in the maintenance of a balanced economy. Arizona's "four Cs"—copper, cotton, citrus, and cattle—were largely resource-based and controlled by forces outside of the state. The continued dominance of high-technology manufacturing in Arizona's industrial mix clearly has led to a more stable economy over the last 12 years. In recent years, Arizona has managed to monitor its economic development and guide its growth in directions that better anticipate future problems.

Research and Development Environment

High-technology companies choose to locate where there is strong research and development activity. The importance of a major research university in such location and relocation efforts is well-known (Dempsey, 1985; Rogers & Larsen, 1984; Wigand, 1985). More specifically, more than 70% of Phoenix-area respondents in a National Science Foundation-sponsored study within the microelectronics industry indicated that the presence or nearness of Arizona State University (ASU) ranked among the top three reasons to locate in the Phoenix area (Larsen, Wigand, & Rogers, 1987). In this sense then, university research activity can serve as a magnet to attract high-technology firms because of the nature of the research and the presence of faculty and staff. University faculty members engaged in research may establish their own companies or private joint ventures with others as a result of their experience. Also industry may work cooperatively with university researchers, funding all or part of their work, supplying facilities and staff or assisting in targeting the research to corporate needs.

Engineering Excellence at
Arizona State University

In 1979, in response to the growing high-technology base forming in the Phoenix area, a 50-member advisory council of engineering was organized at ASU. The council is composed of leaders from high-technology industries in Arizona, and representatives from state government and Arizona State University's College of Engineering and Applied Sciences. Initial goals of the council were to evaluate the ASU engineering program and to develop a strategy for bringing the College of Engineering and Applied Sciences up to a national-leadership status. The advisory council's plan included several five-year phases:

Phase I. This phase called for $32 million in funds to come from state government and private industry. This goal was exceeded by far with $54 million ($12 million in equipment) raised for the Engineering Excellence Program. (Phase II, described below, covers the period of 1985 to 1990.) During this first phase Arizona State University established the Engineering Research Center (120,000 square feet, with a 4,000-square-foot class-100 clean room, portions of which are class 10), including specialization in computer science, transportation, computer-integrated manufacturing, telecommunications, energy systems, solid state electronics, and aerospace.

During this five-year time period, 65 new faculty lines were added (a 59% increase) and 52 new graduate assistant lines (a 334% increase) (Beakley, Backus, & Kelley, 1985). Sponsored research increased to approximately $9.5 million in 1984. Undergraduate enrollment increased 32% from 1979 to 3,351 in 1984; graduate enrollment grew 37% from 1979 to 977 in 1984.

In 1980, a Center for Professional Development was established to meet the increasing demand by engineering and applied science professionals for continuous updating and maintenance of their technical competency and skills. An Interactive Instructional Television Program (IITP) began broadcasting courses to off-campus sites in 1982. Engineering and computer science courses are directed to almost 20 high-technology companies in the Phoenix area. Through a sophisticated teleconferencing system, students at the remote sites are able to interact with the faculty members giving lectures and with students at other remote sites.

Phase II. The Phase I accomplishments have been impressive, and the goals of Phase II (1985-1990) are equally ambitious. According to a 1983 National Academy of Sciences study, the Phase I efforts placed the College of Engineering and Applied Sciences in the top 20 engineering programs within the United States. Phase II plans to raise another $62.5 million for the Engineering Excellence Program to complement the $54 million already invested. Through 1989, various government sources exceeded their goal of $22.5 million by committing more than $49 million. By the end of 1989, a total of $28 million had been received from industry—a figure already far exceeding the Phase II corporate goal of $20 million—in contributions, research contracts, and grants. Total commitments for Phase II are expected to exceed $82 million, nearly $20 million more than the goal.

As an outcome to Phase I and II efforts with industry, corporations within various fields of engineering in the metropolitan Phoenix area recruit 45% to 68% of all new university hires from ASU's College of Engineering and Applied Sciences. This success is not only a local one: Intel nationwide recruits 32% of all new university hires from ASU, exceeding graduates from the University of California at Berkeley.

Phase III. The next five-year phase (after 1990) looks equally promising. The goal is to raise $125 million, $32.5 million of which is projected to come from corporations. At the time of this writing, six corporations have committed themselves for $20 million.

The Arizona State University Research Park and Technology Transfer

Today there are more than 200 university-affiliated research parks in the world, over half of which are located in the United States (C. Boettcher, personal interview, 1989). Ideally, a university-affiliated research park should bring university researchers together with their counterparts in industry to apply research for profit-making ventures. This approach encourages more university research and places the university on the cutting edge of new technological developments. Often universities make it possible for their researchers to share in any financial profits through research-patent incentive and licensing programs. In the past, universities have often been criticized for their "ivory tower" activities and attitudes; this criticism does not hold up as well today—at least for the fields of engineering, business, and applied sciences—as universities have become much more reliant upon interaction with their own communities. The key commodity they have to offer is knowledge and information. Research parks have become a most suitable vehicle to transfer such knowledge and information.

It has been said that a minimally functioning chip-production line costs in the neighborhood of $50 million. To teach a class focusing on such production problems can only be accomplished successfully by professors if they cooperate with industries to gain such knowledge in the first place, as few universities are likely or can afford to make such investments. Cash-strapped universities are having trouble just keeping up with crumbling bricks and mortar. In 1989, a survey by the National Science Foundation found that universities and colleges had identified $11.7 billion worth of renovation or new construction needed for research facilities. The costs associated with such projects are astronomical: A campus complex that will accommodate radioactive materials and toxic gases can cost $20 million to $50 million. Because of budget constraints, the universities planned to spend merely $3.1 billion on such work. In addition, the federal government has been of little help; during the last 20 years there has been no government program to support general research facilities. Although the Academic Research Facilities Modernization Act of 1988 authorized $900 million over five years for repairing and renovating science facilities at universities and non-profit research organizations, so far only $20 million has been approved.

The professor working in such an environment is almost forced to develop a somewhat symbiotic relationship with industry in order to stay up to date in this fast-changing area of research. Without access to such industry facilities, it is becoming increasingly difficult for many universities to continue engaging in leading-edge research. Today, in areas such as microelectronics, it is a given that one needs to have access to modern facilities to carry out research.

The ASU Research Park brings with it the expectation of technology transfer and has three overall goals:

(1) to provide a locational focus for technology transfer between people and programs at the university and their counterparts in private companies engaged in research and development of high-technology products and processes;

(2) to aid in the economic development of the Phoenix area and of the state as a whole; and

(3) to provide an endowment for the university through long-term leasing of building lots.

These three goals are viewed as complementary, as each would be difficult to accomplish without accomplishing the others.

The pressure placed on ASU by industry to improve the engineering program was possibly the prime mover in the development of the research park. The proposal for the park, however, came in 1982 from the president of ASU and two local businessmen. The research park was approved in 1983 by the Arizona Board of Regents. In 1984 the board approved the master plan for the park and authorized the creations of a separate seven-member nonprofit corporation (Price-Elliot Research Park, Inc.) that now manages and markets the park.

On December 4, 1984, ground was broken on 323 acres of land located in southeast metropolitan Phoenix, within the city of Tempe. Skyharbor International Airport is just 15 minutes away by car, and ASU is merely five miles away. A total of $40 million was invested in land and an infrastructure for the park. Individual building sites on 206 acres and one multitenant building (the Transamerica Research Center)—with a total of 52,000 square feet of space—are available to industry. The park has three major tenants occupying their own buildings and 10 smaller ones located in the Transamerica Research Center. Ultimately, the master plan calls for the development of approximately three million square feet of space. The city of Tempe has improved the feeder roads surrounding the site.

The ASU Research Park offers tennis courts, an FAA-approved heliport, jogging paths, equestrian trails, picnic areas, and three lakes. A conference center and hotel site including a child-care learning center, a health management facility, pool, restaurants, and support services are planned. Although interaction with ASU is not a requirement for park tenants, to date nearly all tenants have some programs underway with the university and all have contributed in some fashion to regional economic development. The park's major tenant in terms of laboratory space and number of employees is Fiberite Composite Materials, a division of ICI Composites, Inc., which located in the park in 1987. ICI contributed in 1988 toward the emerging microscopy department through ASU's Industrial Associates Program and has designed several research projects to be coordinated with ASU's programs. ICI employees have been using ASU's new microscope for its materials group's research. ICI also has assisted ASU in its search for experienced composite materials staff and is offering its own staff for guest-lecturer status with various departments in the College of Engineering and Applied Sciences. ICI intends to relocate several more of its divisions from around the world and develop on the park's site its own 40-acre worldwide research and development campus. The most recent transfer in this effort has been ICI's Ceramics Research Group; almost immediately after this group's director (Brian Starling) arrived, he sponsored a two-day symposium between ICI and ASU scientists to develop mutual interests and cooperative interactions for the future direction of both parties. A two-year plan was developed that makes possible the better coordination of various technical issues benefiting both ICI and ASU.

One other technology transfer case deserves to be described and probably has exemplified best what the park's visionaries and founders originally had in mind. VLSI Technology Inc., a recent transplant from Silicon Valley, arrived at the Research Park in 1987. VLSI almost immediately began working with two ASU professors (Lex Akers and Bob Grandin) in the area of solid state electronics. VLSI donated its newly developed integrated-semiconductor chip design software to ASU for the school's Hewlett-Packard workstations. VLSI engineers provide instruction to ASU students and faculty for the new design software. The company sponsors also a cooperative program through which a top ASU student works at VLSI's facility in the research park during the last six months of his or her senior year. Because these students have already been familiar with VLSI's design program for several years, they can be productive designers starting day one. ASU in this process acquires the software design tools and instruction, timely and relevant projects from real-world applied projects can be worked on, and VLSI offers job opportunities and valuable work experiences.

Intel Corporation soon will be purchasing and remarketing two highly complex integrated circuit-chip sets produced by VLSI Technology, Inc. The chips, to be used in IBM AT-compatible personal computer systems, were designed at VLSI's facility in the ASU Research Park and will allow personal computer makers who employ Intel's 32-bit 386 microprocessors to reduce the number of logic chips used in machines from 25 to fewer than 6. The result will be a personal computer that costs less to build and is smaller than many current models. The silicon for the chips will be processed in San Jose, California, and San Antonio, Texas, with final product testing taking place in the VLSI facility in the ASU Research Park.

Interest in SemiMAC, the Arizona-based semiconductor manufacturing consortium, has been substantial. In all, 17 companies have signed on to help fund the activity, and negotiations are continuing with 30 other firms. SemiMAC wants to develop a unique, almost automated semiconductor production concept called "microfactories." These small, computer-controlled silicon processing plants could be cloned easily at locations around the country, enabling small electronics companies to produce their own complex chips in small quantities. They also could be located at major universities to provide advanced semiconductor-engineering laboratories. SemiMAC soon will be moving from its current office (located within the ASU College of Engineering and Applied Sciences) to its own 52,000-square-foot building at the ASU Research Park; the research park itself is contributing two acres as a building site.

The ASU Research Park is still young and has some way to go to mature. The barely 5-year-old park, however, is by comparison ahead of many other U.S. university-affiliated research parks. It took Research Triangle Park 7 years to launch its first tenant, and there were 10 years between Stanford Research Park's first and second buildings.

A Study of Technology Transfer in Microelectronics at ASU

Background and Questions

Traditionally, technology transfer has been conceptualized as the movement of hardware or objects between individuals or groups. In recent years, however, research and development activities increasingly have incorporated software processes, as well as other often nonobservable components, into their products. In turn, then, technology transfer has evolved to include the exchange of *research findings* or *information* as types of technology. The transfer of information among potential users thus adds to the older notion of the transfer of objects as the focus of study.

For the present purposes (i.e., a description of microelectronics technology transfer between ASU and the microelectronics industry

within the metropolitan Phoenix area), technology transfer is defined
as the process through which the results from basic and applied research
are communicated to potential users. Such technology, in the form of
technological information emerging from research, flows from research
universities to industry and from industry to research universities as a
two-way information exchange. It should be noted that during such
processes, cooperation between industry and universities can be encouraged by supportive governmental organizations and programs.

The purpose of the present National Science Foundation-funded
study (see also, Larsen, Wigand, & Rogers, 1987) was to gain an improved understanding of the nature of industry-university cooperative
research and of the technology transfer process in microelectronics. As
already indicated above, this study involved an in-depth investigation
of technology transfer and cooperative research activities between one
research university, ASU, and surrounding microelectronics companies
in the Phoenix metropolitan area. Three broad questions guided the
research:

(1) What are the perceptions of cooperative research and technology transfer
    held by industry and university researchers?
(2) What are the characteristics of cooperative industry-university research
    and technology transfer?
(3) What are the characteristics of cooperative industry-university research
    and technology transfer relationships?

Method

Five study groups included university officials, university researchers, industry officials, industry researchers, and government officials.
These groups were chosen because each represented a somewhat different perspective on industry-university relationships. Industry and
university personnel constitute the core of the industry-university relationships, with government officials encouraging and supporting the
interaction.

Five structured interview questionnaires were developed, one for
each group of respondents. The questionnaires contained items addressing activities and external factors influencing technology transfer.
Questions covered various aspects of microelectronics research, including how contact between an industry researcher and a university researcher first was established, activities involved in developing and

conducting cooperative research projects, reactions to and outcomes of cooperative research projects, and transfer of findings from cooperative research projects to potential users. A total of 56 structured interviews were conducted, involving the following participants: 12 university officials, 14 university researchers, 10 industry officials, 11 industry researchers, and 9 public officials.

Following the interviews, tapes of respondents' comments were transcribed and coding categories developed to allow for quantification of responses to each question. The structured nature of the interviews lent itself well to this procedure. A single rater coded all interviews, transferring the information from the transcripts to computer coding sheets. A random sample of interviews was recorded later by a second rater as a check; no coding differences between the two raters were found. The interview data also were analyzed for findings using a domain-analysis procedure.

Results

The following are generalized results pertaining to the initiation of cooperative research activity, the conduct of cooperative research activity, and effects of cooperative research, as well as technology transfer.

1. Initiation of Cooperative Research Activity

A. *Initial contact.* Industry and university researchers agreed that industry usually was responsible for initiating cooperative research in microelectronics. Industry researchers usually established contact with researchers in a university (58% of the cases), determined the nature of the research project (55%), and planned the research activities (55%). Industry researchers stated that in those cases when industry did not initiate contact unilaterally, industry researchers worked jointly with their university colleagues to begin the research project.

B. *Strategies for making contacts.* Researchers in industry and researchers in the university used different strategies to make contact with their counterparts. Industry researchers relied on their own personal knowledge and experience as the main strategy for identifying potential research partners in 36% of the cases. A second strategy, also used by industry researchers in 36% of the cases, was attending presentations made by university-based researchers at conferences or reading researcher's publications. In contrast, university researchers relied on their colleagues to recommend potential research collaborators in 61% of the cases.

University researchers used established networks to identify and gain access to research partners. Industry researchers relied on their own personal experiences with outstanding researchers. Industry researchers also depended on traditional methods of publications and presentations reporting current research findings; university researchers did not use this strategy.

C. Time required for initiating research. Cooperative research usually involved only experienced researchers. In more than 70% of the cases, industry researchers and their university colleagues both reported that they had known each other—or known about each other—for three to five years before they worked together on a project. Because contacts were formed over a period of years, only those researchers with experience had the necessary backgrounds to conduct cooperative industry-university research.

2. Conducting the Cooperative Research Activity

A. Research topic. Researchers in both locations agreed that modeling and design were the most common topics of cooperative research projects in microelectronics. Industry researchers were not always pleased at this limited focus, but the two groups were in agreement that cooperative research usually meant work in model building or testing.

B. Type of research activity. Researchers in industry and researchers in the university agreed that for them, cooperative research usually consisted of consulting. University researchers reported that cooperative research took the form of consulting in 79% of the cases, while industry researchers reported consulting was their predominant cooperative research activity in 46% of the cases. Some industry researchers reported that their research involvement included other activities, especially faculty appointments for predetermined periods of time.

C. Defining research goals. Industry researchers unilaterally determined the nature of the research half of the time; university researchers cooperated with industry researchers in defining the research goals in other cases. There were no cases reported in which university researchers defined the research goals alone.

D. Planning and conducting the research. Following the identification of research goals, the next step was to plan specific research activities. Industry representatives stated that in 55% of the cases one party planned the specific activities. In contrast, university researchers reported that in 79% of the cases, research activities were planned jointly. University participants reported that they played a more active role in research planning than industry staff acknowledged. Both industry and university researchers agreed that the objectives of the research project usually were specific rather than general. Apparently, cooperative research usually addressed objectives that were well defined so that staff had clear ideas of what was expected.

E. Resources. Cooperative research projects typically had the effect of expanding the availability of resources for research. Both groups of researchers stated that cooperative research activities improved the availability of research equipment and laboratories. University researchers perceived a much greater impact from cooperative research than industry researchers on the availability of equipment. Both groups of respondents reported that the availability of personnel was a major resource provided by cooperative research. University researchers also cited increased access to their own colleagues in the university, noting that cooperative research activities facilitated access to students and other faculty, and also improved communication among departments.

F. Funding. Industry and university researchers had different opinions on the funding sources for their research projects. Industry researchers reported that in 56% of the cases the cooperative research projects were funded jointly by government and industry, and that the remaining 44% of cooperative research projects were funded by industry alone. University respondents perceived that the research was supported by industry in 93% of the cases.

G. Time required. Industry researchers more often perceived cooperative research to be long-term, while university researchers perceived the research to be project specific. Furthermore, 62% of university researchers reported that the cooperative research activity lasted less than one year, while 66% of the industry researchers reported that the research continued over several years.

3. Effects of Cooperative Research

A. Overall satisfaction. Researchers in both locations reported very high rates of satisfaction with cooperative research projects; 90% of industry researchers and 79% of university researchers were satisfied with their cooperative research experiences. Only 10% of industry researchers and 7% of university researchers were dissatisfied. Researchers from industry and researchers from the university judged that both organizations generally were willing to participate in cooperative research; however, their willingness was not overwhelming. University researchers believed that the university would like to conduct cooperative research with industries, but that university administration did not want to provide too many concessions or too many changes in order to do so.

B. Benefits. The greatest benefit of cooperative research for industry researchers was access to university faculty and students; 70% of the industry researchers agreed that this result was most important. The greatest benefit perceived by university researchers was access to industry funds, equipment, and knowledge. Other benefits that attracted university researchers were improved communication with colleagues and access to students.

C. Costs. There were costs associated with cooperative research projects, but these were mentioned less frequently by industry researchers than by the university. The greatest cost of cooperative research for researchers in industry was the complexity of administration; the greatest cost for university researchers was travel time.

4. Technology Transfer

Researchers from industry and the university both acknowledged that technology transfer occurred as a result of cooperative research, and stated that it should occur. Almost all researchers stressed that an exchange of ideas and technical information occurred in cooperative research. University researchers did not perceive industry-university interactions as producing a one-way information flow, with all information going to the university, or with all information going to industry. Rather, researchers noted a two-way exchange of information.

Researchers differed in their judgments of the types of information that should be transferred. Industry researchers stated that findings from basic research were the type of information most appropriate for transfer. University researchers felt that detailed research results were most subject to transfer.

Researchers also cited a number of additional benefits from the cooperative interaction, including enhanced personal knowledge. Researchers in the university used this knowledge to help graduate students become aware of current issues in industry.

Technology transfer appears to have had the greatest effect on researchers who were involved directly in the projects with more limited effects on their nonparticipating colleagues. Researchers agreed that in their experience it was rare for reports of research findings to be presented directly to colleagues who did not participate in the research. Rather, the benefits of cooperative research generally aided those involved directly in the research by increasing their knowledge of a topic, increasing awareness of industry or industry concerns, and providing contacts for student employment.

Discussion

Much research in microelectronics today is conducted in industry rather than in universities. Historically, universities have been responsible for most of the nation's basic research activity in a variety of scientific specialties, while industry concentrated on applied research. Industry now conducts more basic microelectronics research, as well as most of the applied research. The emergence of industry as the prime locus of research activity occurred at the same time that university-based research suffered from funding cutbacks and lack of resources.

Universities continue to be active participants in microelectronics research, in spite of constraints caused by limited funds. University personnel—faculty, graduate students, research associates, and other staff—conduct research in university laboratories and research centers. University-based researchers also cooperate with their industry-based colleagues to address issues of common interest. The research topic often provides a bond connecting the two researchers; however, the researchers do not necessarily share the same opinions about the research project, nor do they necessarily share a common philosophy toward research in general.

Several characteristics of cooperative industry-university research in microelectronics were identified by the present research. Cooperative research projects usually were initiated by industry. Industry generally was interested in working with university researchers because: (1) a corporation had a specific problem and needed help in addressing it; or (2) corporate personnel shortages created a need for additional staff to work on pressing research issues.

The present finding reporting industry initiation of cooperative research differs from results reported in a study of industry-university cooperative research projects conducted by the National Science Foundation (Johnson & Tornatzky, 1984). The NSF study found that university participants played more of a leadership role in initiating the projects and in performing the research tasks. In fact, university impetus in project initiation was an important correlate of favorable project outcomes. By contrast, the present study found that industry provided the impetus for cooperative research projects in microelectronics, and that industry played a key role in orienting the nature of the research activity.

One explanation for the difference in findings may be the researchers and research activity involved. The NSF study investigated cooperative research programs that were funded by the NSF and addressed a variety of scientific and technical areas. NSF staff were often crucial in brokering the relationship between the investigators (Johnson & Tornatzky, 1984). The present study investigated cooperative research projects that developed between industry and university researchers without the support of any third party and that were funded by industry (although government funds may have been awarded to industry previously). Furthermore, the present study was limited to an investigation of research in microelectronics, and did not include cooperative research in other areas as the NSF study did.

Results from the present study and the NSF study agree in finding that cooperative research projects often rely on existing relationships between researchers in industry and researchers in a university. Prior relationships between researchers were common, with consulting relationships being particularly important. In addition, both studies found that researchers who participated in cooperative research were experienced and relatively senior.

Cooperative research projects typically have the effect of expanding the availability of resources for research. Researchers in industry and researchers in the university stated that cooperative research improved their access to research equipment and laboratories. Both groups reported that the availability of personnel also was a major resource.

Researchers in the present study and researchers in the NSF study agreed that they generally were satisfied with their cooperative research participation. Findings from the present study indicate that the greatest benefit of cooperative research for industry researchers was increasing their access to university faculty and students. The greatest benefit for university researchers was access to industry funds, equipment, and knowledge. The greatest cost of cooperative research—one identified by both groups of researchers—was the obstacles created by administrative red tape.

Cooperative research, by its very nature, leads to technology transfer. The purpose of cooperative research is to combine the knowledge and skills of people from different organizations, in this case industry and university, to address a common topic. When researchers from different organizations coordinate their efforts, knowledge possessed by one person flows to others. Likewise, findings emerging from the research activities of one individual pass to colleagues working on the same topic in other locations. If the researchers were not supportive of transferring information, they probably would not participate in cooperative research projects, but would conduct their research in the proprietary settings of their own corporate or university laboratories. Technology transfer—in this case, the transfer of technological knowledge—is therefore an outcome inherent in cooperative research.

Cooperative research encourages the transfer of research results between the university and industry. Although greater contact between organizations helps technology transfer, differences in the purpose and philosophy of the two institutions can create barriers to cooperation. For industry, technology transfer raises the issue of corporate benefit versus national well-being. Microelectronics companies must guard the status of their project development efforts in order to survive and prosper. Consequently, microelectronics companies are reluctant to open proprietary development programs to others, including colleagues in other companies or in universities who may later collaborate with potential competitors.

The traditional role of the university is to educate students and to conduct basic research, making the results available to all. Most research in industry is conducted to advanced product design or production, and the results of the research are used for the corporation's benefit. The incentive for industry researchers is to address issues of interest to the corporation, keeping the research findings proprietary.

Most researchers are aware of the conflicting practices and expectations that exist between themselves and their colleagues in industry or academia. Consequently, researchers working in cooperative ventures attempt to bridge the gap between the institutions, while respecting the differences that exist. The university-based researcher brings to industry the university's assets—a long-term, more theoretically oriented perspective. The industry-based researchers brings to the university industry's assets—state-of-the-art applications and knowledge of future trends. Often the assets of the other group are perceived as detriments; therefore, it becomes the job of the researchers to convince their colleagues in a language that they can understand that these qualities are valuable. Ideally, cooperative research brings the strengths of both perspectives to bear on a common problem.

A promising indicator of technology transfer resulting from cooperative research is the development of networks between university researchers and industry researchers in microelectronics (see also, for example, Wigand & Frankwick, 1989). Networks help researchers develop communication links that connect them with the most current findings and with appropriate sources of information. Networks also introduce a growing number of industry-based researchers to the potential of the university. Likewise, university researchers become more familiar with the needs and problems of industry. As each group comes to know the other better, each is able to use the other as a resource in addressing common topics.

Virtually all parts of the country—as well as other nations—are interested in attracting high-technology industry, and many state and regional programs are directed toward achieving that goal. Officials of microelectronics companies in the present study were divided in their perceptions of the importance of state and local government development efforts on location decisions. Half reported that state and local programs had an effect, and half reported that such programs did not enter into their decisions. When asked whether the state development policies had influenced their own location decisions, only one in five reported that such policies were important in attracting them to their eventual locations.

By contrast, the role played by universities was much more critical to corporate location decisions; 70% of these corporate officials reported that the presence of strong university-industry interaction was a great attraction to them in considering alternative locations. The remainder said that industry-university interaction was only somewhat important in selecting a location.

In those cases where high-technology development programs have been successful, multiple factors have been involved. The community's recent history of public and private development activities is the most important (Chmura, 1982). Past efforts provide experience that leads to understanding and trust among the various elements of a community or region. Social and economic conditions also influence high-technology development. Other factors influencing successful high-technology development include the presence of a research university, a skilled labor pool, and available financing.

Cooperative research benefits both industry and the university, although in different ways. Universities serve industry by providing well-prepared graduates. Industry serves as an alternative source of funding for universities conducting research in microelectronics; however, industry does not replace government as a supporter of basic research. The research activity supported by industry is limited and deals with applied topics that have implications for product development. Industry-supported research less frequently addresses basic research issues. For this reason, industry-supported cooperative research is not a substitute for broadly supported research that produces knowledge of a more far-reaching nature.

References

Arizona Office for Economic Planning and Development (OEPAD). (1983). *Arizona horizons: A strategy for future economic growth.* Phoenix, AZ: OEPAD.

Beakley, G. C., Backus, C. E., & Kelley, R. W. (1985). *Results of the first five-year phase: Excellence in engineering for the 80s.* Report prepared by the Dean's Advisory Council. Tempe, AZ: College of Engineering and Applied Sciences, Arizona State University.

Bloch, E. (1986, May 2). Basic research and economic health: The coming challenge. *Science,* 595-599.

Boettcher, C. [Executive Director of the International Association of University-Affiliated Research Parks]. Personal interview. November 6, 1989.

Bureau of the Census. (1989, November 22). *U.S. Department of Commerce News* (p. 2). Washington, DC: Department of Commerce.

Chmura, T. (1982). *Redefining partnership—Developing public/private approaches to community problem solving: A guide for local officials.* Menlo Park, CA: SRI International.

Dempsey, K. (1985). High-tech race on for silicon areas in U.S. *Plants, Sites & Parks, 12,* 1, 3, 17, 20-22.

Fowler, D. R. (1984). University-industry research relationships. *Research Management,* 27(1), 35-41.

Johnson, E. C., & Tornatzky, L. G. (1984). *Cooperative science: A national study of university and industry researchers.* Washington, DC: National Science Foundation.

Larsen, J. K., & Wigand, R. T. (1987). Industry-university technology transfer in micro-electronics. *Policy Studies Review, 6*(3), 584-595.

Larsen, J. K., Wigand, R. T., & Rogers, E. M. (1987, February). *Industry-university technology transfer in microelectronics.* Report submitted to the National Science Foundation.

Rogers, E. M., & Larsen, J. K. (1984). *Silicon Valley fever.* New York: Basic Books.

Snyder, D. R., & Blevins, D. E. (1986). Business and university technical research cooperation: Some important issues. *Journal of Product Innovation Management, 3*, 136-144.

Wigand, R. T. (1985, August). *High technology development in Arizona.* Working paper, Arizona State University.

Wigand, R. T. (1988). High technology development in the Phoenix area: Taming the desert. In R. W. Smilor, G. Kozmetsky, & D. V. Gibson (Eds.), *Creating the technopolis: Linking technology commercialization and economic development* (pp. 185-202). Cambridge, MA: Ballinger.

Wigand, R. T., & Frankwick, G. L. (1989). Interorganizational communication and technology transfer: Industry-government-university linkages. *International Journal of Technology Management, 4*(1), 63-76.

7

New Business Ventures: The Spin-Out Process

GLENN B. DIETRICH
DAVID V. GIBSON

New business ventures are important as: (1) magnets for research funds; (2) a means of assisting local and state governments in the economic development of a region; (3) a way to capitalize on research activities by commercializing technology; and (4) a means of measuring the success of research and development programs (SRI International Report on Revitalization at Regional and Local Levels, 1984). There is considerable interest in understanding how the research activities of industries, federal laboratories, and universities enhance new company formation through the spin-out process.

Rostow (1987) emphasizes that there has been a shift in international power away from military capability toward economic competitiveness. This shift is illustrated by the rise in the economic power of Japan and other Pacific Rim countries. If the United States is to remain competitive in the world economy, then partnerships between the federal government, universities, and industry must become more effective in linking scientific research and the commercialization of technology through the spin-out process. Glenn B. Dietrich is Assistant Professor of Management and Information Science in the College of Business at the University of Texas at San Antonio. The institutional affiliation of David V. Gibson is described in the introductory chapter of this book.

The Role of Federal Laboratories

Over the decade of the 1980s, policy decisions by the federal government have sought to promote U.S. industrial commercialization of research results from government-funded laboratories (see Appendix A in Chapter 5). The foundation for progress in the area of federal support of technology development and transfer was laid during the Carter Administration (Reams, 1986). The Stevenson-Wydler Technology Innovation Act of 1980 signaled a new governmental concern by injecting market influences into the federal laboratory system. The Bayh-Dole Act of 1983 encouraged increased innovative activity and technology transfer by allowing units cooperating in federally funded research and development to apply for patent rights on discoveries financed through governmental funding. These two laws stipulate that federal research should be made available to industry, and that federal institutions would be held accountable for the transfer of research with commercial potential. These laws also permit intellectual property rights or royalties to be turned over to inventors, who may start companies to exploit their discoveries.

The Small Business Innovation Research Act of 1982 required agencies with large research and development budgets to establish small-business innovation research programs. The National Cooperative Research Act of 1984 further clarified the legal status of joint research and development ventures and encouraged their formation. An executive order titled *Facilitating Access to Science and Technology* was released in 1987 and is intended to promote the dissemination of discoveries of federally funded research programs. Although these governmental actions promote the diffusion of innovation and the promotion of small businesses, they do not address the mechanisms that are required to transfer the technology and to establish a spin-out company (Rahm, Bozeman, & Crow, 1988).

Factors that make for a successful spin-out or new product introduction are discussed in the diffusion literature, where the roles of national laboratories, the diffusion process, and spin-out companies are closely related (Obermayer, 1982; Rahm et al., 1988). One of the main factors cited as necessary for supporting spin-out companies is the requirement for national laboratories to provide reports describing their research and making these reports available to business firms. These reports are supposed to include all of the necessary drawings and process descriptions that would facilitate a spin-out decision (Olken, 1983; Weiss, 1985). However helpful they may be, though, such reports do not necessarily precipitate the start-up of a new firm or the marketing of a new product. Furthermore, the distribution of the federal reports is limited despite formal procedures to make the research results available to the general public (Weiss, 1985).

The Argonne National Laboratory

The Argonne National Laboratory illustrates how important technology transfer has become to federal laboratories, their funding agencies, and their clients. Largely as a result of the Stevenson-Wydler Technology Innovation Act and the Bayh-Dole Act, this laboratory actively promotes the commercialization of technology and participates in the Federal Laboratory Consortium Resource Directory, which provides a nationwide directory of federal resources that private industry can check for the latest information concerning federal research efforts.

The Argonne Technology Transfer Center was formed in 1985 with the goal of transferring technology to industry and of making industry aware of Argonne's expertise, patents, and inventions. Because of the center's tie to the University of Chicago, much research conducted at the university is included in the Argonne effort. In 1986, 37 patents authored by Argonne and University of Chicago engineers were described in brochures for prospective licensers and venture capitalists (Caruana, 1987).

The Role of Industry

Along with federal programs that are becoming more proactive in promoting technology commercialization through spin-out companies, U.S. industry is also increasingly promoting cooperative effort to strengthen technology transfer (Bopp, 1988). According to Rogers and Larsen (1984), organizational concepts such as providing the atmosphere to promote innovation, developing rewards for risk taking, improving communication lines, and setting up transfer mechanisms are all necessary if the firm desires to promote the spin-out process.

Although corporations need to manage developed activities to ensure their continuation, new ventures require management that encourages change and the development of new ideas (Mendell & Ennis, 1985). New ventures have four special characteristics that require special management styles: (1) uncertainty, (2) knowledge intensiveness, (3) competition with alternate methods, and (4) the need for interdisciplinary and interfunctional information exchange. Management that allows a free flow of information, creates teamwork across divisions, and encourages identification with company goals will help foster innovations and speed the establishment of new ventures (Kanter, 1985).

Essentially every business organization has the potential for being a business incubator that provides the training and experience necessary for a successful business start and shares the risks associated with a new business venture. According to studies conducted by Cooper (1971), the characteristics of the incubator organization appear to influence the ways in which entrepreneurs leave the parent company to start their own companies—and, to some degree, these same characteristics influence the spin-out companies' subsequent success.

New spin-out firms usually locate in the same geographic region as the parent organization where the founder(s) worked (Cooper, 1971; Cooper & Dunkelberg, 1981b; Susbauer, 1972). The percentage of founders who remain in the same geographic region in which they lived prior to the start of the new company was 75% for a broad study of 890 founders across the United States (Cooper & Dunkelberg, 1981b). In 1984, Cooper (1985a) analyzed 59 new firms that were characterized as being growth oriented and capable of having a substantial economic impact. The sample was divided into four categories: (1) electronics and computers; (2) software; (3) biotechnology; and (4) nontechnical firms. The firms had median sales of $42 million, and 84.9% of them in electronics and computers, software, and nontechnical areas started in the same geographic area as the incubator organization. And 57% of the biotechnology sample located in the same area.

By staying in the same geographic region as the parent company, spin-out personnel can utilize existing networks of business associates. Such location also allows entrepreneurs to remain employed in their current jobs so that the spin-outs can be part-time efforts until they are established firmly. Furthermore, potential family problems associated with moving—due to leaving friends and neighborhoods, and changing schools—are also eliminated.

Studies by Cooper and Dunkelberg (1981a,b) have found that spin-outs were related closely to the technology of the incubator organizations and the work that the founders did at the incubator organizations. This is especially the case in high-technology spin-outs where current knowledge of technologies and markets is essential for success (Cooper, 1985a). Accordingly, the spin-out may be in competition with the incubator organization and may foster a conflict of interest for the founder in terms of trade secrets and business clients.

Spin-out companies frequently are founded by teams of two or more people. In a study of 955 high-technology firms, Shapero (1971) found that 59% of the foundings involved teams. The teams permit more well-rounded groups to form and build the companies. For example, a strong researcher can combine with someone who is skilled in the business aspects of running a company.

Research suggests that substantial numbers of spin-out companies from established firms are the result of strong negative push factors exerted upon the founders (Shapero & Sokol, 1982). People who are frustrated with their employment or who have difficulty getting along with their co-workers tend to look for other work opportunities. People who have their ideas for new products or processes rejected or who see a technology-based opportunity that their company does not want to pursue are likely candidates to champion spin-out activities. On the other hand, Dietrich (1990) found that there were no significant push factors that encouraged the spin-out process from the university setting.

The Role of Universities

The first U.S. land-grant universities were established in 1855 because of the perceived need for institutions of higher learning to become more involved in social and economic development. The Morrill Act in 1862 gave large tracts of land as an endowment for agricultural colleges and defined the service mission for land-grant institutions. The Hatch Act in 1887 initiated the trend toward transferring technology from the universities by appropriating $15,000 for each state to be used for establishing experimental stations that would have the obligation of transferring university technology. The Smith-Lever Act in 1914 provided funding for the land-grant colleges and stated that part of their mission would be to conduct a program of cooperative extension work in the agriculture and mechanical sciences through instruction and demonstrations to individuals outside of the universities. Thus, the traditional role of the state university has been to educate students as well as to conduct basic and applied research.

During World War I, U.S. universities and the federal government joined forces to conduct military research. The Overman Act allowed the federal government to create laboratory structures involving the military and selected universities for the purpose of doing research to assist the war effort (Collins & Tillman, 1988). After the war, the laboratories were turned over to the universities. During World War II, universities and private industry both were encouraged to conduct military research and to build on the laboratory structures associated with the World War I research effort (Collins & Tillman, 1988). The most notable of these partnership efforts was the Manhattan Project, which included the Department of Defense, the University of Chicago, and Oak Ridge National Laboratory. This collaboration developed the first atomic bomb.

Since World War II, there have been other periods of close cooperation in research and its commercial application between the federal government, universities, and industry. For example, the Atomic Energy Commission provided research funds to many universities to study atomic physics. Much of this research has been used in both military and civilian products. The launching of Sputnik in 1957 precipitated an unprecedented U.S. cooperative effort that tied together science, research, and technology. Research conducted at universities is credited with developing the electronic computer and exploring the electronic properties of semiconductors (Collins & Tillman, 1988).

Prior to 1980, individual federal agencies were allowed to formulate their own policies concerning ownership of intellectual property rights that resulted from federally funded research. Under these conditions universities were reluctant to commit time or resources without knowing whether they would own the resulting innovations. The Patent and Trademark Amendments Act passed in 1980 gave the ownership of intellectual property rights resulting from federally funded research to the universities that conducted the research (Erickson & Baldwin, 1988). Between 1980 and 1985, universities surveyed by the General

TABLE 7.1: Participation in Research and Development by Industry, Federal Government, and Universities as of Mid-1980s

	Percentage of Research Conducted	Percentage of Total Research Funds
Industry	73	50
Federal Government	12	47
Universities	9	2

SOURCE: Lindsey, 1985

Accounting Office experienced a 74% increase in business sponsorship of university research.

The Importance of Basic Research

As Table 7.1 shows, universities performed about 9% of the total research and development conducted in the United States in the early 1980s (Lindsey, 1985). In constant 1972 dollars, the expenditures for research and development were about the same in 1985 as they were in 1967, so the amount of money going to universities to fund federal research programs has been declining in real terms (Lindsey, 1985).

When research is categorized into basic, applied, and developmental forms, the importance of universities to basic research is illustrated (Table 7.2). Even though universities conduct only about 9% of the total research in the United States, they conduct 57% of all basic research (Lindsey, 1985). Of the total research and development that a university participates in, 67% is basic, 27% is applied, and 6% is developmental. In contrast, industry spends about 4% of its research dollars on basic research, 20% on applied, and 76% on developmental (see also Abetti, LeMaistre, & Wacholder, 1988; Botkin, 1988; Dempsey, 1985; Rogers & Larsen, 1984; Ryans & Shanklin, 1988).

In the late 1970s, the growth curve of cooperative activity between business and higher education began to increase significantly in the types and numbers of partnerships (Powers, Powers, Betz, & Aslanian, 1988). According to studies conducted at the Battelle Memorial Institute, as global economic competitive pressures increased and as government research funding declined in the 1980s, universities began to promote the commercialization of technologies (Steinnes, 1987; Wilem, 1990). Funds provided by industry for university research and

TABLE 7.2: Research Allocations by Industrial, Governmental, and University Sectors as of Mid-1980s (by percentage)

	Basic Research	Applied Research	Developmental Research
Industry	4	20	76
Federal Government	14	22	64
Universities	67	27	6

SOURCE: Lindsey, 1985

development programs have increased from 3% in 1981 to 5% in 1985 (Lindsey, 1985).

Although there is documentation that industry- and government-sponsored basic research conducted by faculty at research universities is important to the economic well-being of the United States (Powers et al., 1988), basic research alone does not promote the spin-out process nor the transfer of technology. The 1982 report of the Joint Economic Committee of Congress stated that 67% of high-technology companies surveyed relied on student recruiting as the most important means of technology transfer, 46% considered university publications important, and 42% thought that government distribution of basic research results was important. The support of university basic research was considered an important means of technology transfer by only 25% of the respondents. Because other factors such as information dissemination, entrepreneurism, and capitalization are involved, technology transfer does not necessarily follow investment in basic research (National Science Foundation, 1983).

The University and the Spin-Out Process

The extent to which universities function as incubators of spin-out companies varies considerably. In the areas around Palo Alto, California; Boston, Massachusetts; Austin, Texas; and Ann Arbor, Michigan, many of the new technology-based firms are direct spin-outs from a university or from one of the research laboratories associated with a university (Dietrich, 1990; Lamont, 1972; Roberts, 1972; Susbauer, 1972). Although all major universities conduct research and educate students, there are few successful business spin-outs in the vicinity of these universities. Proximity to a good university seems to be a necessary but not sufficient condition for development of high-technology spin-out companies (W. Brown, 1985).

Cooper (1971) analyzed spin-out companies in Palo Alto, California, that were begun during the 1960s and found that Stanford University did not play a major direct role in spinning out these high-technology companies. Cooper concluded, however, that universities have played a major indirect role in the spin-out process by attracting entrepreneurs to an area, educating them, and adding to the overall quality of life and general attractiveness of an area—thereby encouraging the entrepreneurs to remain in the area after their education was completed. Assistance in solving technical problems and providing continuing education also were cited as indirect influences in the decision to start a spin-out company close to a university. Cooper's (1971) findings concerning the relatively passive role of Stanford University in the spin-out process during the 1960s were largely replicated by Dietrich (1990) in his study of the role of the University of Texas at Austin in the spin-out process from 1955 to 1988.

Although spin-out activity has not been traditionally one of the primary tasks of the university professors or staff (T. Brown, 1985), as national and international business environments consider spin-out activity more important, the operating paradigms as defined by some universities are being modified to value and reward applied research and technology transfer operations associated with the founding of new firms (McQueen & Wallmark, 1985, 1988).

There are three main ways that universities encourage the spin-out process: (1) sell the rights to the innovation; (2) license the rights to the innovation; and (3) arrange for an equity position in the venture (Botkin, 1988; Giannisis, Willis, & Maher, 1988; Wigand, 1988; Wilson & Szgenda, 1990). Some universities have initiated formal arrangements with "technology transfer companies"; these companies provide links between the university-based technology and the marketplace. The companies license the innovations from the universities and are then responsible for the commercialization of the ventures. The company makes the determination as to whether or not the new products or processes have the potential for being profitable. License fees generally are shared between the universities and academics who conducted the research.

Universities also establish their own technology licensing offices and support networks to promote the commercialization of academic research (McQueen & Wallmark, 1985; Morrison & Wetzel, 1990; Wilson

& Szgenda, 1990). According to Doutriaux (1987), many Canadian universities have technology transfer offices whose primary purpose is to manage the universities' research efforts and to facilitate interaction with industry. These offices apply for patents, arrange licensing agreements, and work with the venture-capital industry for new product start-ups of university-based research. Such offices frequently assist the faculties in establishing ventures.

Although the faculty of a university may have sophisticated technical knowledge (Mark, 1988), they often do not have (or want, for that matter) experience selling ideas, establishing firms, or managing businesses. The incubator organization can assist in these endeavors as well as provide the environment within which the founders can form effective, multidisciplined teams (Hisrich & Smilor, 1988; Smilor, 1986).

The University of Texas at Austin and Texas A & M University both provide examples of universities establishing offices to facilitate the technology transfer/spin-out process. At the University of Texas, the Graduate School of Business and the IC^2 Institute have cooperated to form the Center for Technology Venturing, which includes the Austin Technology Incubator (ATI). The ATI is designed to nurture and accelerate the development of new technology-based firms by providing support in such key areas as a low-overhead environment and the use of shared resources and highly qualified professionals in the functional areas of business and engineering. The ATI is funded jointly by the city of Austin, the Austin Chamber of Commerce, private sources, and the University of Texas at Austin and thus represents a true consortium of public and private interests.

At Texas A & M University, the Texas Engineering Experiment Station has established the Office of Technology Business Development (TBD). Part of the mission of TBD is the commercialization of research results that offer excellent commercial potential, with an emphasis on university-developed technology. TBD assists the start-up company with a variety of services such as commercial feasibility studies, market research, business plan development, product development assistance, intellectual property information, and information on financing alternatives. These services are provided at various centers throughout the state. In essence, TBD serves as a network to link the faculty researchers with the business community and provides business and technical assistance to the developing firm.

From the university's perspective, private ventures have been an important source of funds especially in times when federal grants for research and enrollments in graduate schools are down (Doctors, 1969). In 1987, licensing fees amounted to $6.1 million for Stanford University, $5.4 million for the University of Wisconsin, and $3.1 million for Massachusetts Institute of Technology (Jereski, 1988).

In 1988, U.S. universities spent about $13 billion in research and development (R&D) activities, and about $45 million was collected in royalties. (Of the approximately $120 billion spent by federal laboratories in R&D activities only about $4.4 million was collected in royalties.) MIT collected a higher amount of royalties on its R&D than most other U.S. universities: $6.2 million in royalties on $700 million in research. As of early 1990, MIT receives about 1 to 2 university-based inventions per day and the university processes about 3 to 4 patents per week. MIT alumni-founded companies provide about $10 billion in annual income to Massachusetts and over 300,000 jobs (J. Preston, personal communication, 1990).

Also important are the nonfinancial benefits of university spin-out activities. Spin-out companies can provide a dynamic atmosphere for promoting cutting-edge research and the application of research results, a positive influence on teaching, an exciting environment for attracting and retaining quality graduate students and faculty, and an increase in the perceived positive presence of the university in the community, as well as enhanced national and international prestige (Doutriaux, 1987; McQueen & Wallmark, 1985, 1988; Rogers & Larsen, 1984).

Public/Private Strategic Alliances to Promote Spin-Off Activity

As new kinds of institutional developments among business, government, and academia are beginning to promote economic development and technology diversification (Merrifield, 1987; Ryans & Shanklin, 1986; Sexton & Smilor, 1986), universities are becoming more proactive in the spawning of new businesses (Brett, Gibson, & Smilor, 1990). The initial mission of the land-grant universities to provide service and disseminate information in the agricultural and mechanical sciences is being expanded to include all forms of technology transfer. The university increasingly is being looked upon to become a generator for cutting-edge technologies and economic development in addition to the more traditional roles of teaching and research (Smilor, Gibson, & Kozmetsky, 1988).

The emerging role of universities—which includes an emphasis on developmental research and on fostering an environment necessary to promote high-technology spin-out companies—often is met with the concern that university-based commercial activities and interests may interfere with fundamental university values (Allen & Norling, 1990). Although many leading universities have long-established paradigms that emphasize basic research and teaching as their primary focus and have developed reward and motivation systems that reinforce these behavior patterns, other excellent universities (such as Stanford, MIT, the University of Texas at Austin, and the Rensselaer Polytechnic Institute) have developed environments that encourage entrepreneurial behavior in addition to traditional university activities (Smilor, Kozmetsky, & Gibson, 1988).

Although in most U.S. universities the spin-out process has not been managed proactively (Smilor, Gibson, & Dietrich, 1990), the successful introduction of new products and the formation of spin-out companies increasingly requires the cooperation of government, industry, and university. Langfitt, Hackney, Fishman, and Glowasky (1983) developed a model of the technology production cycle (Figure 7.1). This model illustrates the importance of the linkages among public and private entities in promoting economic growth through basic research and technological innovation. The model illustrates how knowledge either leaves the system or becomes applied research and, eventually, the basis for new products or processes. One of the primary challenges of the spin-out process shown in the Langfitt model is technology transfer. It is not sufficient that basic research take place and that it generate knowledge, but the technology must be put into a marketable form so that it can contribute to the economy.

The Case of Tracor, Inc.

The roles of the federal government, business, the research university, and the influence of entrepreneurial role models on spin-out activity can be effectively demonstrated through a case study of Tracor, Inc., a high-technology company located in Austin, Texas. Frank McBee, the founder of Tracor, earned both bachelor's (in 1947) and master's (in 1950) degrees in mechanical engineering at the University of Texas at Austin after serving as an Army Air Corps Engineer from 1943 to 1946. He decided to stay in Austin to raise his family because of the high

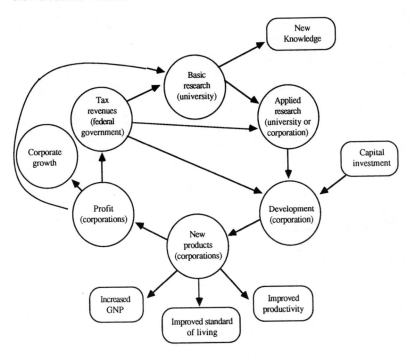

Figure 7.1. Linkages Between Government, Industry, and Academia in Promoting Economic Growth Through Basic Research and Technological Innovation

SOURCE: Langfitt, Hackney, Fishman, & Glowasky (1983), p. 178.

quality of life in the area. In the late 1940s, McBee became an instructor and then an assistant professor at the University of Texas in the Department of Mechanical Engineering. In 1950, he became the Supervisor at the Defense Research Laboratory (now called the Applied Research Laboratory) at the University of Texas' Balcones Research Park, which had been associated with World War II research.

In 1955, McBee joined forces with three physicists from the University of Texas to form Associated Consultants and Engineers, Inc., an engineering and consulting firm. Drawing on their training and work experience, the four scientists focused their efforts on acoustics research and were awarded a $5,000 federal research contract for an industrial noise-reduction project. In 1957, they formed Texas Research Associates (TRA). During the late 1950s and early 1960s, the scientists

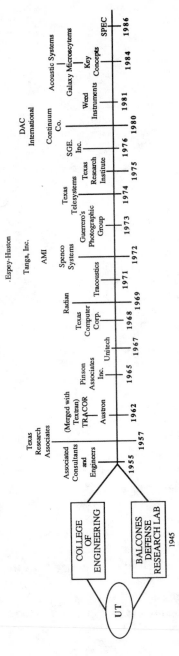

Figure 7.2. The Formation and Spin-Outs from Tracor

SOURCE: Smilor, Kozmetsky, & Gibson, 1988.

taught and did research at the University of Texas at Austin and at Balcones Research Park while working on developing TRA.

Figure 7.2 shows the stream of educated talent and technology from the College of Engineering and the Defense Research Laboratory at the University of Texas at Austin that formed the entrepreneurial venture of Associate Consultants and Engineers, which led to the establishment of Tracor in 1962. Tracor was successful and grew rapidly to become Austin's only homegrown *Fortune 500* company. Even more impressive, however, is the constant stream of entrepreneurial talent that has spun out *from* Tracor. More than 20 companies have spun out of Tracor and have located in Austin since 1962 (Figure 7.2); these companies also created spin-outs of their own. Radian Corporation, for example, has spun out four companies: Nova Graphics, BPI, ZYCOR, and Meister Engineering. Some of these spin-outs from their parent companies were "friendly," and some were not. Nevertheless, Figure 7.3 dramatically shows the job-creation impact of Tracor and its spin-outs on the Austin area. As of 1985, a total of 5,467 persons were employed in these spin-outs and their respective offspring.

The life cycle of the birth of Tracor (the child of the university and federally funded research) and the spin-out companies from Tracor (i.e., the university's grandchildren) and the spin-outs from the spin-outs (i.e., the university's great-grandchildren) came full circle in 1990 when the leveraged buy-out of Tracor (which loaded $700 million of debt on the company in 1987 and which totaled $747 million in November, 1989) and the subsequent cutbacks in the U.S. defense industry led to a downturn for the parent company. Many of the spin-outs of Tracor, however, and the spin-outs of the spin-outs continue to grow and prosper.

Through the years since its founding in 1962, Tracor and its spin-outs have provided a broad range of dividends to the University of Texas, the city of Austin, the state of Texas, and U.S. international competitiveness. These dividends have been in the form of employment (both college-educated and less formally skilled workers); local and national taxes; fostering a regional environment conducive to high-tech economic growth and entrepreneurism; contributing to the University of Texas in many financial and professional ways; and in contributing to U.S. industrial competitiveness.

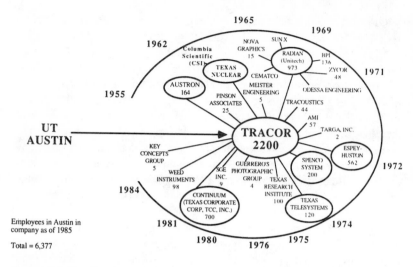

Figure 7.3. The Employment Impact of Tracor and Its Spin-Outs

SOURCE: Smilor, Kozmetsky, & Gibson, 1988.

Conclusion

Cooperation among U.S. universities, federal laboratories, and businesses is becoming increasingly crucial to the timely use and commercialization of new technology wherever it is developed. As a result, U.S. public and private institutions are becoming more proactive in the promotion, transfer, and commercialization of technology. This cooperation is perceived as an important means of accelerating technology transfer and the commercializing processes to reinvigorate the U.S. economy and to sustain a high quality of life.

References

Abetti, P. A., LeMaistre, C., & Wacholder, M. (1988). The role of Rensselear Polytechnic Institute: Technopolis development in a mature industrial area. In R. Smilor, G. Kozmetsky, & D. Gibson (Eds.), *Creating the technopolis: Linking technology commercialization and economic development.* Cambridge, MA: Ballinger.

Allen, D. N., & Norling, F. (1990). Exploring perceived threats of faculty commercialization of research. In A. Brett, D. Gibson, & R. Smilor (Eds.), *University spin-out companies: Economic development, faculty entrepreneurs, and technology transfer.* Totowa, NJ: Rowman & Littlefield.

Bopp, G. (Ed.). (1988). *Federal lab technology transfer.* New York: Praeger.

Botkin, J. W. (1988). Route 128: Its history and development. In R. Smilor, G. Kozmetsky, & D. Gibson (Eds.), *Creating the technopolis: Linking technology commercialization and economic development.* Cambridge, MA: Ballinger.

Brett, A., Gibson, D., & Smilor, R. (Eds.). (1990). *University spin-out companies: Economic development, faculty entrepreneurs, and technology transfer.* Totowa, NJ: Rowman & Littlefield.

Brown, T. L. (1985, Fall). University-industry research relations: Is there a conflict? *Journal of the Society of Research Administrators, 17*(2), 7-17.

Brown, W. S. (1985). A proposed mechanism for commercializing university technology. In J. A. Hornaday, F. A. Tardley, Jr., J. A. Timmons, & K. Vesper (Eds.), *Frontiers of entrepreneurship research.* Wellesley, MA: Babson College, Center for Entrepreneurial Studies.

Caruana, C. (1987, August). Technology transfer at Argonne National Laboratory. *Chemical Engineering Progress,* 94-98.

Collins, T. C., & Tillman, S. A., IV. (1988). Global technology diffusion and the American research university. In J. T. Kenny (Ed.), *Research administration and technology transfer.* San Francisco, CA: Jossey-Bass.

Cooper, A. (1971). *The founding of technologically-based firms.* Milwaukee, WI: Center for Technology Management.

Cooper, A. C. (1985). Contrasts in the role of incubator organizations in the founding of growth-oriented firms. In J. A. Hornaday, F. A. Tardley, Jr., J. A. Timmons, & K. Vesper (Eds.), *Frontiers of entrepreneurship research.* Wellesley, MA: Babson College, Center for Entrepreneurial Studies.

Cooper, A. C., & Dunkelberg, W. C. (1981, August). *Influences upon entrepreneurship—a large scale study.* Paper presented at the Academy of Management Meetings, San Diego, CA.

Dempsey, K. (1985, March-April). High-tech race on for silicon areas in U.S. *Plants, Sites, & Parks, 12,* 1, 3, 7, 20-22.

Dietrich, G. B. (1990). *Toward a new university paradigm: A perspective from technology start-up companies.* Doctoral dissertation, University of Texas at Austin.

Doctors, S. I. (1969). *The role of federal agencies in technology transfer.* Cambridge: MIT Press.

Doutriaux, J. (1987, Fall). Growth pattern of academic entrepreneurial firms. *Journal of Business Venturing, 2*(4), 285-297.

Erickson, A., G., & Baldwin, D. R. (1988). The new frontier of technology transfer. In J. T. Kenney (Ed.), *Research administration and technology transfer.* San Francisco, CA: Jossey-Bass.

Giannisis, D., Willis, R., & Maher, H. (1988, April). *Technology commercialization in Illinois.* Paper presented in conference at the University Spin-Off Corporation, Virginia Polytechnic Institute and State University, Blacksburg.

Hisrich, R. D., & Smilor, R. W. (1988, Fall). The university and business incubation: Technology transfer through entrepreneurial development. *Journal of Technology Transfer, 13*(1), 14-19.

Jereski, L. (1988, May 2). Patent profit. *Forbes,* 104.

Kanter, R. (1985, Winter). Supporting innovation and venture development in established companies. *Journal of Business Venturing, 1,* 45-60.

Lamont, L. M. (1972). The role of marketing in technical entrepreneurship. In A. Cooper & J. Komives (Eds.), *Technical entrepreneurship: A symposium*. Milwaukee, WI: Center for Venture Management.

Langfitt, T. W., Hackney, S., Fishman, A. P., & Glowasky, A. W. (Eds.). (1983). *Partners in the research enterprise*. Philadelphia: University of Pennsylvania Press.

Lindsey, Q. W. (1985, Fall). Industry-university research cooperation: The state government role. *Journal of the Society of Research Administrators*, 85-90.

Mark, H. (1988). Basic science and the technopolis. In R. Smilor, G. Kozmetsky, & D. Gibson (Eds.), *Creating the technopolis: Linking technology commercialization and economic development* (205-208). Cambridge, MA: Ballinger.

McQueen, D. H., & Wallmark, J. T. (1985). Innovation output and academic performance. In J. A. Hornaday, F. A. Tardley, Jr., J. A. Timmons, & K. Vesper (Eds.), *Frontiers of entrepreneurship research*. Wellesley, MA: Babson College, Center for Entrepreneurial Studies.

McQueen, D. H., & Wallmark, J. T. (1988). *University technical innovation: Patents, spin-offs and academic research*. Paper presented in conference at the University Spin-Off Corporation, Virginia Polytechnic Institute and State University, Blacksburg.

Mendell, S., & Ennis, D. (1985, May-June). Looking at innovation strategies. *Research Management, 28*, 33-41.

Merrifield, B. D. (1987, Fall). New business incubators. *Journal of Business Venturing, 2*(4), 277-284.

Morrison, J., & Wetzel, W. (1990). Support network for faculty spin-off companies. In A. Brett, D. Gibson, & R. Smilor (Eds.), *University spin-out companies: Economic development, faculty entrepreneurs, and technology transfer*. Totowa, NJ: Rowman & Littlefield.

National Science Foundation. (1983). *University-industry research relationships: Selected studies*. Washington, DC: National Science Foundation.

Obermayer, J. H. (1982). Government R&D funding and startups. In K. Vesper (Ed.), *Frontiers of entrepreneurship research*. Wellesley, MA: Babson College, Center for Entrepreneurial Studies.

Olken, H. (1983, November). Cooperation between private industry and the national laboratories. *Proceedings 1983, IEEE Conference on Engineering Management*. Dayton, OH: IEEE.

Powers, D. R., Powers, M. F., Betz, F., & Aslanian, C. B. (1988). *Higher education in partnership with industry*. San Francisco: Jossey-Bass.

Rahm, D., Bozeman, B., & Crow, M. (1988, November-December). Domestic technology transfer and competitiveness: An empirical assessment of roles of universities and governmental R&D laboratories. *Public Administration Review, 48*(6), 969-978.

Reams, B. D., Jr. (1986). *University-industry partnerships: The major legal issues in research and development agreements*. Westport, CT: Quorum.

Roberts, E. B. (1972). Influences upon performance of new technical enterprises. In A. Cooper & J. Komives (Eds.), *Technical entrepreneurship: A symposium*. Milwaukee, WI: Center for Venture Management.

Rogers, E. M., & Larsen, J. K. (1984). *Silicon Valley fever: Growth of high-technology culture*. New York: Basic Books.

Rostow, W. W. (1987). On ending the cold war. *Foreign Affairs, 64*(4), 831-851.

Ryans, J. K., & Shanklin, W. L. (1986). *Guide to marketing for economic development.* Columbus, OH: Publishing Horizons.

Ryans, J. K., & Shanklin, W. L. (1988). Implementing a high tech center strategy: The marketing program. In R. Smilor, G. Kozmetsky, & D. Gibson (Eds.), *Creating the technopolis: Linking technology commercialization and economic development.* Cambridge, MA: Ballinger.

Sexton, D. L., & Smilor, R. (Eds.). (1986). *The art and science of entrepreneurship.* Cambridge, MA: Ballinger.

Shapero, A. (1971). *An action program for entrepreneurship.* Austin, TX: Multidisciplinary Research, Inc.

Shapero, A., & Sokol, L. (1982). The social dimensions of entrepreneurship. In C. Kent, D. Sexton, & K. Vesper (Eds.), *Encyclopedia of entrepreneurship.* Englewood Cliffs, NJ: Prentice-Hall.

Smilor, R. (1986, May-June). Commercializing technology through new business incubators. *Research Management, 30,* (36-42).

Smilor, R., Gibson, D. V., & Dietrich, G. B. (1990). University spin-out companies: Technology start-ups from UT Austin. *Journal of Business Venturing, 5,* 1-14.

Smilor, R., Gibson, D., & Kozmetsky, G. (1988). Creating and sustaining the technopolis: High technology development in Austin, Texas. *Journal of Business Venturing, 4,* 49-67.

Smilor, R., Kozmetsky, G., & Gibson, D. (Eds.). (1988). *Creating the technopolis: Linking technology commercialization and economic development.* Cambridge, MA: Ballinger.

SRI International. (1984). *Economic revitalization and technological change: Next steps for action and research, summary report.* Menlo Park, CA: SRI International.

Steinnes, D. N. (1987, August). On understanding and evaluating the university's evolving economic development policy. *Economic Development Quarterly, 1*(3), 214-225.

Susbauer, J. C. (1972). The technical entrepreneurship process in Austin, Texas. In A. Cooper & J. Komives (Eds.), *Technical entrepreneurship: A symposium.* Milwaukee, WI: Center for Venture Management.

Weiss, G. (1985, July 22). Selling science—universities try to turn a profit from research. *Barrons, 20,* 28.

Wigand, R. T. (1988). High technology development in the Phoenix area: Taming the desert. In R. Smilor, G. Kozmetsky, & D. Gibson (Eds.), *Creating the technopolis: Linking technology commercialization and economic development.* Cambridge, MA: Ballinger.

Wilem, F. (1990). Forming spin-off corporations through university/industry partnerships. In A. Brett, D. Gibson, & R. Smilor (Eds.), *University spin-out companies: Economic development, faculty entrepreneurs, and technology transfer.* Totowa, NJ: Rowman & Littlefield.

Wilson, M., & Szgenda, S. (1990). Promoting spin-off companies through university equity participation. In A. Brett, D. Gibson, & R. Smilor (Eds.), *University spin-out companies: Economic development, faculty entrepreneurs, and technology transfer.* Totowa, NJ: Rowman & Littlefield.

8

Transfer via Telecommunications: Networking Scientists and Industry

FREDERICK WILLIAMS
ELOISE BRACKENRIDGE

Dating from the 1960s, when scientists first gained remote access via telecommunications networks to mainframe computers to share programs, files, or leave messages for one another, we have seen the evolution of networks for resource sharing and communications. Such network services are playing an increasingly important role in technology transfer, not only in supporting collaboration among scientists and giving them access to a wide variety of computing and data base resources, but also in linking them to government and commercial groups charged with bringing technology to the marketplace. In this chapter the authors review examples of such networks and services, then conclude with a view toward priorities for further development. Frederick Williams, whose background was summarized in the introductory chapter of this volume, is Director of the Center for Research on Communication Technology and Society at the University of Texas at Austin. Eloise Brackenridge, at the time of this writing, was completing a doctorate at that institution. Her background includes executive positions in communications, planning, and marketing with large multinational corporations. Earlier in her career she served overseas with the U.S. Department of State's Foreign Service.

Telecommunications and Technology Transfer

The process of technology transfer is enhanced greatly by the voice, data, and video capabilities made available with modern telecommunication networks. Through a world wired with copper (twisted pair), coaxial, and fiber optic cable—supported by satellite, radio, and microwave transmission—researchers are finding solutions by using their own telephones, terminals, personal computers, or workstations to access an enormous array of resources. Telecommunication networks enable geographically dispersed researchers to consult and confer in real time with colleagues around the world. These networks provide the link to vast libraries of data and to the powerful computation capabilities of supercomputers. Telecommunication is the revolutionary research tool that has extended our eyes, ears, and minds into a new universe of networked experimentation and investigation.

Researchers are able to leverage their resources by sharing expensive equipment, such as radio telescopes, with an entire "invisible college" (researchers all investigating a particular school of thought). Obscure information once stored within the confines of a single isolated library is now available on-line and worldwide to all who are connected to the network. Simulations and calculations of inaccessible phenomena such as the process of superconductivity or the collisions of neutron stars, requiring the fastest supercomputers, are now available to academic, industrial, and governmental scientists.

These are only the first steps in technology transfer, however. The next part of this expansion is to link university, governmental, or commercial offices concerned with technology transfer as a function into the network. This may be a newly organized university office charged with promoting commercialization of campus laboratory inventions where the network provides mail services to faculty scientists, assists in creating proposals, or offers a data base that tracks university research activities. A city or state government office may use the same linkages to promote its operation, while also using the network to stay in touch with federal laboratories or the agencies publishing requests for proposals (RFPs).

A research consortium might link itself to all participating groups, scientists, government offices, and its own sponsoring organizations or shareholders. A shareholder company may see a market opening that could be communicated on the network to a consortium whose staff may in turn use the network to scan university laboratories for relevant research activities.

Technology Transfer

In examining telecommunication networks and their role in technology transfer it is important to consider the nature of the transfer. The process is often very circuitous, however, with ideas and notions moving in and out of abstraction and definition in the course of developing an innovation. Typically, many communications on the same or similar topic will occur. Telecommunication will not assist in making this course more linear and sequential, although a few projects will appear to evolve this way. The large majority of technology transfer projects will exhibit some apparent chaos and disorder, with multiple iterations during the discovery and development process. What telecommunication networks can do for this abstract and disordered process is to speed up, intensify, and help clarify aspects of the unfolding project. What networks can do extremely well (in conjunction with computers) is keep a myriad of participants informed and up-to-date, organize a vast amount of overwhelming data, provide instantaneous communication with significant parties regardless of their geographic location, and provide access to advanced services such as supercomputing and parallel processing.

The Network Infrastructure

In most applications the *traditional network* refers to the integrated international, national, and local exchange telephone networks originally designed for voice but now experiencing rapid growth in data communication services. In the United States this infrastructure—once the domain of the national monopoly, American Telephone and Telegraph System—now is serviced by telecommunication companies such as the long-distance carriers MCI and US Sprint, and the local exchange carriers owned by the regional Bell holding companies such as Southwestern Bell and U.S. West, or by many independent companies, of which GTE is a major player.

We use these networks every day to speak on the telephone or to send a facsimile message, a TWX, a Telex, a video image, or a computer signal. These are called *common carriers* because they are a public network presumably available to all, but are supplemented by private networks either constructed by or for special users, as well as virtually private networks that are leased services from the public network. Many of the telecommunications services discussed in this chapter are served by common carriers, sometimes in combination with private networks.

A wide variety of specialized telecommunication networks can be accessed by universities, industrial research facilities, and government laboratories or offices. Roughly speaking, there is an informal hierarchy of networks, descending in size from the overarching common-carrier networks which carry voice and data around the globe, to the specialized international data networks, to the regional or wide area networks (WANS), to local area networks (LANS).

Computer Networks

Computer networks are essentially telecommunications systems (transmission facilities, switches) dedicated to linking users to computers and computers to one another. The most frequent and well-known application is for user log-on from terminals, workstations (or personal computers equipped for terminal emulation) to computers that provide computation, data base, or communications services. It is increasingly common for such services to involve logging into a network of computers through a "switch," then connecting users to the computer selected for the particular application. It is this network-switching or gateway function that is particularly important in technology transfers because it supports such important applications as electronic mail and messaging, file transfer, and access to data bases ranging from specialized bibliographic services to electronic journals.

The above services often are divided into two broad categories: computer-mediated communication, and resource-sharing services. Either of these may be delayed (batch) or immediate (interactive). Table 8.1 summaries some of the principal uses of network services.

TABLE 8.1: Services Often Provided by Computer Networks

TYPICAL BASIC SERVICES

> Electronic mail (messages from one person to another)
> Bulletins (messages from one person to many)
> News (messages from many to many)
> Conferences (conversations between a few or many)
> File transfer (movement of computer files from a source)
> Computation
> Data base access and search function
> Encryption (encoding) and security protection (passwords)
> Directory service (similar to white page telephone book)
> Gateways or interconnections to other networks or data bases

ADVANCED SERVICES

> Video conferencing (compressed video or full scan video, one-way, two-way,
>   multiways with video picture or audio-only response)
> Computer graphics (computer aided design, modeling, and other advanced
>   graphics in color)
> Parallel computation requiring parallel computers
> High-level computation requiring supercomputers
> Simulation (e.g., weather, astronomy) requiring supercomputers

Computer-Mediated Communication and Resource-Sharing Services

Batch computer-communication services generally include one-to-one (mail), one-to-many (bulletin boards), and many-to-many (news). Interactive computer communication services are instantaneous and immediate with no delay between messages. By contrast, resource-sharing services are based on remote connection (remote log-on or the ability to access another computer as if logged onto it locally) and file transfer. Other services can include interconnections (gateways) to commercial or international networks, videotext (an interactive system for accessing a wide variety of data bases), and user support. Resource sharing will eventually involve group word processing, editing, and publication services, with participation by various users of the network regardless of their geographic location.

Today the primary challenge of large computer networks is to combat problems with incompatibility in order to create connectivity among many diverse smaller networks. A multitude of protocols (using, for example, IBM or DEC), machine architectures, operating speeds, and capabilities all serve to create a nightmare of incompatibilities that isolate groups of users from one another. Major research networks can be likened to a geodesic dome that permits interconnection between these groups of users running on different standards.

Networks in the 1990s

A number of new applications will be available over the next several years to assist researchers and others with technology transfer. As networks move to broader bands and faster speeds, new kinds of services will come available: The network of the future will be digital, integrated (carrying audio, text, and video), and completely interactive.

Also to be anticipated as networks become more "intelligent" are new kinds of software for analysis such as *expert systems* that use facts and rules of judgment, expert decision making, and logic to find lines of reasoning which lead to answers or solutions. An expert system incorporates much more than textbook or manual knowledge. It is able to mimic the basic reasoning process of the human expert upon whom the system is patterned (Feigenbaum, McCorduck, & Nii, 1988). In essence, a key function of an expert system is efficient technology transfer: storing the vast knowledge of an expert person, thereby capturing a body of experience and expertise and passing it on to others who may be separated by both time and distance. These systems can be used to diagnose a broad spectrum of conditions ranging from human illness to manufacturing backlog. They also are employed to advise, to plan, and to provide specialized knowledge.

Research Networks

Especially important for technology transfer are the computer networks typically found in use in major research universities. These range from a wide variety of local area networks to memberships (or linkages into) interuniversity networks like BITNET (and now Internet), and larger scale high-capacity networks like NSFNET. These networks permit researchers at colleges and universities—and industrial and government laboratories—to work simultaneously on a common body of knowledge. This increases the productivity of scholars, hastens the pace of scientific discovery, and accelerates the transfer of research results (Hazleton, 1989). Roberts (1988) describes national and international networking goals:

- Increase research productivity by improving access to information, to supercomputers, and to other specialized sources.
- Advance the quality of academic research and instruction by expanding opportunities for collaboration and cooperation.
- Shorten the time required to transmit basic research results from campuses to the private sector and thus enhance national research and product development capabilities.
- Broaden the distribution of scholarly opportunity and creativity by connecting faculty, students, and staff from diverse geographies.

Internet

Internet—a worldwide network of hundreds (at least 700) of separate local, regional, national, and international networks—links 500,000 users at university, industry, and government research sites. The network is expanding at an enormous rate: 20,000 computers connected in 1987 grew to 60,000 in 1988. Internet acts both as a facility to share resources between organizations and as a test bed for new innovations in networking. There are two Internet network groups based on two different protocols: Internet, which uses the TCP/IP protocol, and DECnet Internet, which uses Digital Equipment Corporation protocols. Many institutions belong to both (LaQuey, 1989; Quarterman, 1990).

The oldest network in Internet is ARPANET, the U.S. Department of Defense Advance Research Project Agency's (DARPA) experimental packet-switched network, introduced in 1969. ARPANET is phasing out in 1990; its traffic will be carried by MILNET and NSFNET (National Science Foundation Network). NSFNET, one of Internet's major constituents, is now and will become even more important to advanced research activities—consequently, this network was chosen from dozens of research-oriented networks to illustrate technology transfer in action in this chapter.

A Close-Up of NSFNET

NSFNET is a national network created to improve communications, collaboration, and resource sharing in the science and research commu-

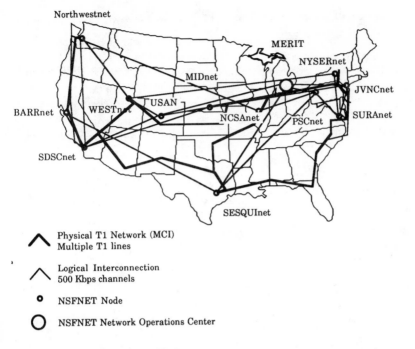

Figure 8.1. NSFNET: Major Nodes

nity. NSFNET interconnects major regional networks at 250 universities and research sites (Wulf, 1989) and provides any recipient of an NSF grant with access to a supercomputer (see Figure 8.1). The network is structured on three levels:

- backbone supercomputer sites consisting of six NSF-sponsored supercomputer centers interconnected on long-haul, very high-capacity trunk lines;
- mid-level networks connected to the backbone and to other mid-level networks and international networks; and
- campus networks.

The NSFNET backbone uses MCI's fiber optic circuit and digital microwave-radio network to carry data. The management and operation of NSFNET's backbone is supervised by MERIT, Inc., a nonprofit consortium composed of eight Michigan universities (LaQuey, 1989).

Close-Up of a Backbone Network: NYSERNet

One of the six major centers forming the backbone of NSFNET is the New York State Education and Research Network (NYSERNet), a high-speed data communications network linking universities, industrial research laboratories, and government facilities in the state of New York. The goal is to give greater access to "computing and information resources which will aid in improving economic competitiveness" (NYSERNet, 1987). NYSERNet is a nonprofit company formed in 1985 by a group of New York educators, researchers, industrialists, and NSF. Among the 47 users of NYSERNet are Columbia University, New York University, Polytechnic University, Cornell University, IBM, Kodak, and Brookhaven National Laboratory. The Cornell National Supercomputer Center and the NorthEast Parallel Architecture Center supply electronic library access and additional network services to NYSERNet (Quarterman, 1990).

NYSERNet was the first network of its type to involve facilities owned by telephone companies, in this case, NYTel and Rochester Telephone. By 1989, however, all the facilities were owned, operated, and maintained by NYSERNet, which leases lines from the common carriers as would any other customer. NYSERNet's revenue stream is 10% supported by grants from the state of New York and the National Science Foundation. The remaining 90% is funded by users of NYSERNet, who pay annual fees from $15,000 to $78,000 (depending upon speed) to access the system.

Projects promoting technology transfer are daily occurrences at NYSERNet. Four examples of technology transfer are the Apple Computer-University of Rochester connection; the Hartford Graduate Center-NorthEast Parallel Architecture Center connection; the Alfred University-Cornell supercomputer connection; and the State University of New York at Buffalo-Ames Research Center connection.

Engineers at Apple Computer in Cupertino, California, conduct a sizable amount of development work in conjunction with the University of Rochester. Using BARRnet to connect to the Internet—and, in turn, to connect to NYSERNet—Apple's researchers are able to do industrially relevant work sitting at their own computers in California.

The Hartford Graduate Center currently is involved deeply in researching artificial intelligence. Researchers desiring to test human depth perception for application to robotics devised a model of how a part of the human brain sorted out information. To test the theory, the center needed a method to mimic the brain that could be provided by parallel processing. Using NYSERNet, the Hartford Graduate Center is able to log in to the NorthEast Parallel Architecture Center to conduct the appropriate tests.

Alfred University is a national center for ceramic engineering where researchers study ceramics for a variety of applications. In order to work with highly complex equations needed for developing models, the university connects with the Cornell National Supercomputer Center through the facilities of NYSERNet (John Eldridge, personal interview, November, 1989).

At the Computational Fluid Dynamics Laboratory at the State University of New York at Buffalo, research in the areas of combustion, computational fluid dynamics, and turbulence is being conducted. Through NYSERNet connections the lab has gained access to supercomputers at Cornell, the University of Illinois, and the NAS-supported CRAY-2 at NASA Ames Research Center (NYSERNet, 1989).

Another kind of technology transfer also can be found on the network. Users around the country have the advantage of public domain software. Users are able to "borrow" very sophisticated (and, otherwise, very expensive) programs whenever they need them by tapping into the network. This method of sharing enables even the smallest-budget researchers to perform at the same level as those who are in big-budget operations.

NSFNET's Mid-Level

Eight mid-level networks are independent entities in a federation linked to the NSFNET backbone. An additional eight networks complete the mid-level group. Among the NSFNET international networks connected through mid-level members are EASINET (European Academic Supercomputer Initiative), JANET (Joint Academic Network, United Kingdom), and JUNET in Japan.

Examples at NSFNET's Mid-Level: BARRNet and THEnet

BARRNet operates on the second level of the NSFNET and connects universities and research organizations in northern California. The network expects to have 40 members at the end of 1989 and more than 60 at the end of 1990. Universities (such as Stanford and four University of California northern campuses) and industrial research members (such as SRI International, Hewlett-Packard, Xerox PARC, and Apple

Computer) share the network with several government and private research laboratories such as Lawrence Livermore National Laboratory, NASA Ames, and Monterey Bay Aquarium Research Institute (Baer, 1989).

THEnet (Texas Higher Education Network) provides service state-wide and to Mexico to 50 members, including universities (such as the 17 campuses of the University of Texas system, Texas A & M, and Rice University), industrial research members (such as Lockheed, Schlumberger, and Texas Instruments) and other organizations such as SEMATECH, Microelectronics and Computer Technology Corporation, and Superconducting Super Collider Laboratory.

THEnet functions in a joint cooperative effort with Sesquinet, a mid-level network connected to the NSFNET backbone. The Texas networks combined form one of the largest of the regional networks and include 2,000 nodes (LaQuey, 1989).

These and other networks will play increasingly important roles in fostering interinstitutional exchange and the diffusion of ideas, products, and innovative procedures. Network building is a dynamic, growing endeavor. Over the years connectivity will improve, and speed and capabilities will increase. Hardware and software issues will be tested with issues of another type—human communication.

Networks as a Research Opportunity

The development of advanced networks can serve a secondary purpose of being an opportunity to study network design, applications, economic impacts, or even user behaviors. One theme of Sheridan Tatsuno's (1986) book on the Japanese plan to use an advanced fiber network to link its "technopolis" cities is that in addition to serving their own communications needs, Japanese scientists and marketing organizations could research the very technology they were using. In the United States, one example comes from New York's Centers for Advanced Technology (CATs), which were created by the state in 1982 with the objective of improving the interface between basic and applied research.

CATs are cooperative research and development centers that bring together New York's state government, its universities, and its private industry. Jointly these groups engage in basic and applied research, with the aim of harnessing new technologies for the economic good of the state. Currently, every center receives up to $1 million annually from the state of New York; these funds are matched by at least an equivalent amount from private industry (New York Director of Economic Development, personal communication). One of the units, the Center for Advanced Technology at Polytechnic University, focuses on telecommunications. For 1986, corporate grants totaled $1.6 million. Some 20 corporations—including IBM, GTE, AT&T, and NYTel—contribute to Polytechnic, and in 1987 the state legislature authorized continued funding at the $1 million level through the 1995 fiscal year (Michael Shimazu, personal communication, March, 1988).

Users and Network Benefits

Beginning with Hiltz's and Turoff's (1978) *The Network Nation*, there has been a growing interest in the social and organizational impacts of telecommunications networks. Innovative programs are being developed by universities and research centers; a small cross-section is mentioned below.

Pennsylvania State University's Distance Learning

Pennsylvania State University, in partnership with the Pennsylvania Educational Communications Systems (PECS), makes televised courses available to more than 700,000 cable subscribers. PECS is a private microwave network built to deliver credit and noncredit courses to subscribers' homes. PECS provide a full-time 24-hour channel through participating cable companies.

This university was the first to experiment with using compressed-video technology to deliver courses. Compressed video maintains a relatively smooth-motion image at speeds 120 times slower than that needed for full-scan video (thus using a fraction of the bandwidth). Interactive compressed video is used to offer courses simultaneously at multiple locations.

The live two-way compressed video is supplemented by one-way full-scan video delivered to all of the university's campuses by satellite. The satellite link-up makes it possible to offer certain courses to all campuses that until recently were unavailable, thereby preventing students from being required to transfer to other locations. Furthermore, Penn State is able to offer a variety of video programming to the campuses and to make its faculty resources accessible to the rest of the world (Augustson, 1988).

Starlink

Starlink is a research network for British astronomers. It provides researchers with interactive computing facilities (hardware and software), including use in image and spectral work. The network helps researchers process data quickly from sources such as satellite astronomy, ground-based telescopes, multiple-dish radio telescope arrays, and automatic scanned photographic plates. According to Quarterman (1990), Starlink has brought together astronomers in the United Kingdom in an integrated community and has vastly increased collaboration and sharing. Starlink has 50 hosts at 19 sites with 950 users in England, Scotland, Wales, and Northern Ireland; 87% of the users are research astronomers.

Online Computer Library Center

Originally incorporated in 1967, the Online Computer Library Center, Inc. (OCLC) services 6,000 university libraries in the United States, Europe, Japan, and Canada. OCLC holds 15 million catalogue records. Each week members add 24,000 new records and OCLC adds an additional 10,000 from organizations such as the Library of Congress and the British Library. OCLC handles 55,000 requests a week through the interlibrary loan system. The center also provides catalogue conversion and reference services on computer disks.

Probably the largest national network of this type, OCLC is a prime example of resource sharing over computer networks. OCLC operates over leased lines or dial-up connections; it also provides gateways to other information services. Additionally, OCLC is involved with a project called Linked Systems Project (LSP)—along with other major centers such as the Research Libraries Group (RLG)—to enhance the easy exchange of bibliographic records and the sharing of authority data (Arms, 1988).

Collaboratories

With a concern for anticipating ways in which a network will be used, the National Science Foundation is spearheading a project called the

National Collaboratory. This is a major coordinated program leading to an electronic "collaboratory" (collaboration + laboratory) or "center without walls" in which all researchers at universities, industrial and governmental laboratories, and consortia can participate regardless of geographic location. Remote interaction is essential because colleagues in an "invisible college" may be widely separated, data may be too vast to store in one location, and some of the most challenging projects may require equipment and facilities that are geographically distant.

Today's scientific challenges focus more and more on remote phenomenon that are difficult to access, distributed across space and time, and are conceptually and computationally complex. The collaboratory requires more than a network itself; it needs a complete infrastructure that provides facilitative software, simulation tools to serve as an analog for a "wet laboratory," remote and interchangeable "smart instruments," digital libraries, and accessible data. The collaboratory will support people-to-people cooperation and collaboration, people-to-instrument interaction (with access to expensive and remote equipment and instruments), and people-to-data interaction (with access to gigantic data bases) (Lederberg & Uncapher, 1989).

The project, in addition to its obvious merits, will enhance research productivity by opening up an untapped pool of research talent at four-year and minority institutions and also provide information that can be translated into products, which will help the country's economy (Wulf, 1989).

There are a variety of services available for use in the collaboratory, including such applications as electronic mail, electronic file transfer, and data base access. Higher-level functions, however, would be important. These could be developed by integrating available technologies with advanced technologies such as:

- digital instrumentation offering real-time control and feedback from remote instruments;
- multimedia meetings offering videoconferencing with high-quality audio and shared computational whiteboards;
- digital mail providing value-added services such as yellow pages and improved addressing mechanisms;
- scientific reference service using human and artificial intelligence; and
- digital journals, peer review, and a digital library.

The long-term development of such a project would require the availability of new and advanced electronic tools not in existence today, including:

- hypermedia conversation support (videoconferencing supported by hypermedia tools such as hypercard and hyperties);
- intelligent agents (such as robots) that can search highly distributed libraries, schedule coordinated experiments, and other activities;
- interoperable data description (data described so that it is understandable to various analysis systems);
- information fusion (techniques for understanding data from heterogeneous sources);
- smart agents for the design of experiments (tools with the "intelligence" to be real "assistants" to scientists and engineers); and
- smart data gathering (incorporates intelligence into the instruments to allow self-directed data gathering).

The collaboratory would be much more than a set of applications or tools for research. It would be a national computer-based infrastructure for scientific research. An important component would be to identify and carry forward "an aggressive rapid-prototyping test bed activity" (Lederberg & Uncapher, 1989).

Telecommunication and Economic Development

There is a growing interest in how research-oriented networks can be seen as contributing in a broad sense to the economic development or competitiveness of a city, state, or region. In two major projects (the former involving the senior author, the latter involving both authors) research application of telecommunications networks was seen as one of the tools for economic development (Schmandt, Williams, & Wilson, 1989; Schmandt, Williams, Wilson, & Strover, in press). An advanced network can assist in:

- providing opportunities for research leading to the marketing of network products (including software) and services;
- attracting R&D or high tech industries;
- hastening the technology transfer process;
- supporting the delivery of continuing education in engineering and the sciences;

- improving library service delivery via such networks; and
- enhancing educational programs in universities and secondary schools, and eventually primary schools.

Toward a U.S. National Network

In the fall of 1989, a plan for a Federal High Performance Computing (HPC) Program was transmitted to Congress by President Bush's Office of Science and Technology Policy; earlier this year U.S. Senator Albert Gore, Jr. introduced legislation proposing the project. Included in this plan is a proposal for a National Research and Education Network (NREN), a federally coordinated government, industry, and university collaboration "to accelerate the development of high-speed computer networks and to accelerate the rate at which high performance computing technologies—both hardware and software—can be developed, commercialized and applied to leading-edge problems of national significance." (D. Allen Bromley in the transmittal letter of the Federal High Performance Computing Program; Executive Office of the President, 1989). The goals of the program are to:

- maintain and extend U.S. leadership in high-performance computing, and encourage U.S. sources of production
- encourage innovation in high-performance computing technologies by increasing their diffusion and assimilation into the U.S. science and engineering communities
- support U.S. economic competitiveness and productivity through greater utilization of networked high performance computing in analysis, design, and manufacturing (Executive Office of the President, 1989).

The HPC program stresses increased cooperation between business, academics, and government in building a network that will serve as a prototype for future commercial networks. In proposing NREN, government sources found that the current national network technology does not support wide-based scientific collaboration or access to unique sources adequately, and often the national networks in the United States stand as barriers to effective high-speed communication. Furthermore, Europe and Japan are moving ahead aggressively in a variety of networking areas, surpassing the current state-of-the-art technology in place in North America.

NREN would be built on the existing infrastructure of long-distance lines and fiber optic cables and would employ new transmission technologies to increase its speed and interconnective ability. NREN's structure, based on the existing informal tier system, would be composed of a federally sponsored "superhighway" providing support for large users and access for every state; a middle tier of regional and state networks with a broadband capability; and a lower level of smaller networks such as local area networks at universities (National Research Council, 1988).

This "supercomputer highway" would connect government, academia, and industry with a network ultimately capable of transmitting 1,000 times more data per second than current networks. Three stages are proposed: In the first stage, the existing Internet (TI) trunk lines will be upgraded to 1.5 megabits per second, a project already under way. As a complement to this, DARPA is undertaking a project called Research Internet Gateway (RIG) to develop "policy-based" routing mechanisms that will allow the interconnection of these trunks. Directory services and security mechanisms also are being added.

In stage 2, upgraded service will be delivered to 200-300 research facilities with a shared backbone network operating at 45 megabits per second. The ability to share this backbone network will reduce costs and improve service. Once the new research backbone is interconnected with the existing NSFNET backbone to and from NREN, it is anticipated that every university and major laboratory will be interconnected.

Stage 3 plans are still being developed, but extremely high-speed 1- to 3-gigabit networks supported by fiber optic trunks will be important. Also targeted are advanced capabilities such as remote interactive graphics, nationwide data files, and high definition television (HDTV).

Comparing the network to the interstate highway system built in the 1950s, the initiative could eventually lead to a "wired nation" where small businesses and homes would share in the capability of HDTV, vast electronic data banks, and other applications.

Concluding Comments

A final and important topic concerns a step beyond extending the network to researchers in remote locations: encouraging them to use the medium for sharing information, education, experimentation, and—most of all—collaboration on new ideas and new products. Empirical evidence has shown that telecommunications research networks are effective for technology transfer, and in numerous instances essential. The days of the solitary inventor have concluded almost completely. Highly motivated information-seeking users have developed new and imaginative ways of communicating with one another electronically; they manage to do so over thousands of miles with electronic mailing lists and news-group bulletins, computer conferences, and information exchanges. Groups such as Starlink (the British astronomy network) and OCEANIC (the ocean research group's data base) have united researchers with common interests. Consider the Swedish participation on NORDUNET; in a period of two years, the Swedish infrastructure grew dramatically to include virtually all higher education institutions and many corporations and government entities, spearheaded by an aggressive national policy.

Of particular interest to communication researchers studying technology transfer are theories of collaboration. Why do some individuals, separated by oceans, cultures, and other barriers such as imperfect software and hardware, persist in collaborating over the networks? Why do other researchers decline?

Part of the answer lies in the essence of the American culture—our spirit of competition. In Europe and Japan collaboration and cooperation on large, expensive, difficult projects are pursued. The thinking behind these projects is that no one entity has the resources in terms of money or staff to undertake the investigation and development of certain products and processes, all of which are largely sophisticated high-tech ventures. To assuage this problem, research groups in universities or in consortia are formed where original research can be conducted and findings can be transferred back to the participating organizations.

Most nations of the developed world lead the United States in numbers and kinds of research consortia and cooperative ventures. Only in 1984 did the government of the United States enable exemption from antitrust action for certain corporate operations; this was late in the evolution of the global marketplace.

It is a difficult action in this country to embrace collaboration. The notion of competition permeates the fabric of our activities—from the playing fields to the boardrooms, strong peer competition is promoted and rewarded by management policy. George Kozmetsky (in Chapter 1 of this book) and others have spoken out about innovative and creative management. They have suggested that there are many new ways to look at leadership and group inspiration. As a first step at fostering technology transfer and cooperation, American supervisors might tackle two platitudes: "not invented here" and "if it ain't broke, don't fix it." The first casts suspicion, even scorn, on any idea that originates from outside an organization. Clearly, if technology transfer is to be successful by word of mouth, by video conference, or by electronic mail, we need to set aside this outdated notion.

The second implores us to maintain the status quo, generally the antithesis of successful technology transfer. Change is risky and new ideas often fail, but it is clear that for the United States to remain economically competitive in the world market—and, in fact, to make headway against the formidable competitors in Europe and the Pacific Rim—we need to begin to look for different models of management and for new tools for investigation, whether offered with a supercomputer or CAD/CAM design.

Successful technology transfer can be enhanced greatly over networks by the participation of strong organizational leaders sharing their vision; by a widening in corporate cultures to embrace innovation and new ideas and to support the incubation of these new processes and inventions; and lastly, by government policy that fosters technology transfer through financial incentives, agency assistance, and other positive methods such as the proposed new high-performance computer highway and the national collaboratory.

The promise of advanced telecommunication is new, as are the methods of collaboration with electronic tools that can span geographic, cultural, and technical distances. These tools hold the potential to accelerate technology transfer and to enhance its efficacy; we have only to look for ways to improve cooperation, to work together to make optimum use of them.

References

Arms, C. (Ed.) (1988). *Campus networking strategies.* Bedford, MA: Digital.

Augustson, J. G. (1988). The Pennsylvania state university. In C. Arms (Ed.), *Campus networking strategies.* Bedford, MA: Digital.

Baer, P. (1989). BARRNet. In T. LaQuey (Ed.), *User's directory of computer networks accessible to the Texas Higher Education Network member institutions.* Bedford, MA: Digital.

Executive Office of the President. (1989). *The federal high performance computing program.* Washington, DC: Office of Science and Technology Policy.

Feigenbaum, E. P. M., & Nii, H. P. (1988). *The rise of the expert company*. New York: Times Books.

Gore, A. (1989, May 19). *National high-performance computer technology act of 1989*. Newsletter.

Hazleton, J. (1989). *A proposal to establish Bluebonnet—a Texas network education and research corporation*. Unpublished manuscript.

Hiltz, S., & Turoff, M. (1978). *The network nation: Human communication via computer*. Reading, MA: Addison-Wesley.

LaQuey, T. L. (1989). *Users' directory of computer networks accessible to the Texas Higher Education Network member institutions*. Bedford, MA: Digital Press.

Lederberg, J., & Uncapher, K. (Co-Chairs). (1989). *Towards a national collaboratory*. Report of an invitational workshop at Rockefeller University.

National Research Council. (1988). *Toward a national research network*. Washington, DC: National Academy Press.

NYSERNet, Inc. (1987). *NYSERNet, Inc.: The New York State Education and Research Network*. Troy, NY: NYSERNet.

NYSERNet, Inc. (1989, July-August). [Entire issue.] *NYSERNet News, 2*(8).

Quarterman, J. S. (1990). *The matrix: Computer networks and conferencing systems worldwide*. Bedford, MA: Digital.

Roberts, M. (1988). Introduction. In C. Arms (Ed.), *Campus networking strategies*. Bedford, MA: Digital.

Schmandt, J., Williams, F., & Wilson, R. H. (Eds.). (1989). *Telecommunication policy and economic development: The new state role*. New York: Praeger.

Schmandt, J., Williams, F., Wilson, R. H., & Strover, S. (Eds.). (in press). *The new urban infrastructure: A study of large telecommunication users*. New York: Praeger.

Tatsuno, S. (1986). *The technopolis strategy*. Englewood Cliffs, NJ: Prentice-Hall.

Wulf, W. A. (1989, Summer). Government's role in the national network. *Educom Review, 24*(2), 22-26.

IV

International Perspectives

9

Mexico and the United States: The Maquiladora Industries

EDUARDO BARRERA
FREDERICK WILLIAMS

Multinational manufacturing is another growing example of the importance of technology transfer—in this case, the transfer of production technology rather than product innovation itself. Such transfer depends upon the ability to decentralize the management system through telecommunications, to "stretch the links" so that design and planning functions (as well as performance feedback) can flow freely between company headquarters and remote manufacturing sites. It also requires that manufacturing sites be linked effectively with suppliers, that fabrication be located—or relocated easily—in favorable labor environments, and that final assembly sites have tariff and transportation benefits derived from physical proximity to markets. The authors have researched this topic in uses of telecommunications to serve the maquiladora (or "twin plants") industries along the U.S. and Mexican border, one of the world's fastest-growing examples of multinational manufacturing. As of this writing, Eduardo Barrera, a native of Mexico,

was completing his doctorate at the University of Texas. He holds
previous degrees from the Monterrey Tech (ITESM) and the Uni-
versity of Texas. He has written about the Mexican telecom-
munications sector for various international conferences. The
background of Dr. Williams, director of the Center for Research
and Communication Technology at the University of Texas at
Austin, was summarized in the introduction chapter to this book.

"Maquiladora" Industries

Development of Production Sharing

Traditionally, the word *maquiladora* referred to the sharing of crops
between the landowner and an itinerant harvester, or in dividing offspring in cattle breeding. In the same production-sharing sense, it refers
to a partnership or division of activity in manufacturing as applied to
the U.S. and Mexican sides of the border. *Maquila* operations have
represented agreements where components are shipped "in bond" (tem-
porarily tax-free) from the United States into Mexico for assembly
by low-cost Mexican labor, then returned with taxes paid only on the
value added in Mexico. The Mexican government has favored maquila
development as it creates jobs for their citizens, and is a mechanism
for technology transfer, as stated explicitly in Article 16 of the 1983
Maquila Industry Decree. The article identifies the aim of the federal government is to "promote investment in advanced technology sectors, and incorporate new technology which modernizes production processes" (Sklair,
1989). U.S. manufacturers cite the advantages of low cost and proximate
Mexican labor over offshore (Asian) alternatives. U.S. border states have
generally countered the "loss of U.S. jobs" argument with the points that
(1) they were already lost to Asia; and (2) that maquilas create product
distribution centers and household shoppers for the U.S. side of the border
economy. *Expansion Management* (1988) cites that 30% of the money
generated from maquilas is spent on the U.S. side of the border. A third
argument is that the maquiladora industry actually creates jobs for
Americans in all states. For example, Mitchell and Vargas (1987)
estimate that it is directly accountable for 175,440 jobs distributed
among all the continental United States. Of the 1,500 maquiladora
plants in Mexico, most are located on the 2,000 miles of the Mexican

border, with large concentrations in Tijuana, Ciudad Juárez, and in the Matamoros-Reynosa area. The state of Nuevo Leon is the one that currently is making the strongest efforts to attract maquiladoras, especially to a new industrial belt that is being master planned and would be an important part of the "Camino Real" project, of which the hub would be the Colombia Bridge west of Laredo. The Camino Real extends from Austin to Saltillo, which includes the high-tech industries of the Austin-San Antonio corridor, an area that would be of strategic importance as one of the main developers and incubators of high technology.

The maquila industry has replaced tourism as Mexico's second primary source of income. Although the $1.7 billion it generates is still much lower than the $9 billion of the oil industry, the economies of an increasing number of families and cities are becoming dependent on maquila operations. The fiscal and legal framework for maquilas in Mexico dates back to 1965, when the government created the North Border Industrialization Program. A recent trend on the capital side of the maquila industry is the increasing number of operations from countries other than the United States. Manufacturing plants of Japanese and Korean capital are mushrooming not only on border cities on the Pacific coast, but all along the U.S.-Mexican border. The advantages the Mexican border offers have resulted in foreign investment that is still increasing.

Growth of Flexible Manufacturing

In recent years, particularly as large companies with maquiladora operations have entered the picture, telecommunications has increased in importance for the efficiency of these operations. This is because the new generation of maquiladoras has many of the characteristics of the new "craft paradigm" described by Piore and Sabel (1984), including:

- *"Just-in-time" supply.* This was introduced in Japan by automobile manufacturers and consists of reducing the parts in stock, thus externalizing the inventory costs.
- *Small batch manufacturing.* This is where the production is more specialized and requires flexible machinery (including robotics) and skilled labor.
- *A decrease in the "Vernon production cycle."* This refers to reducing the time between the different phases from R&D to actual manufacturing.

The generation or exchange of strategic information with regard to R&D becomes more crucial to production activities as firms adopt flexible manufacturing models. In the literature, these models generally are used as synonymous to craft paradigm and to "post-Fordist" production. Patricia Wilson (in press) points out the absence of an autonomous R&D capability in the maquiladora industry, which can be generalized to the whole industrial sector in Mexico. In fact, the policy of the Salinas administration is not to stimulate indigenous R&D, but instead to "acquire, assimilate, adapt and diffuse efficiently the technology" (Poder Ejecutivo Federal, 1989). The maquiladoras have been seen as the potential principal mechanism of technology transfer, an assumption that is reflected in the concessions given to this industry in the Law of Control and Register of Technology Transfer and Use and Exploitation of Patents and Brands (Davis, 1985). Leslie Sklair (1989), however, argues that at the core of the production-sharing strategy is the assumption that knowledge-intensive activities are to remain in the capitalist heartlands, and that a transnational corporation cannot risk losing its share of the production to their Third World partners. Between 1970 and 1987, only 7.2% of the 54,565 patents registered in Mexico were not foreign; The United States accounted for 53.6%, West Germany 9.9%, and other foreign countries 29.2% (Notimex, 1990).

Despite the introduction of microelectronics-based innovations in the equipment of some subsectors (mainly electronics) and the appearance of production cells instead of the long lines of production characteristic of the Fordist factory (Brown & Domínguez, 1989), Wilson (in press) found that flexible automated plants are not growing as fast as those that are using more labor-intensive technology. Sklair (1989) argues that the mere relocation of technology does not result in a genuine technology transfer, which he calls *technology anchorage*.

The characteristics of the maquiladora industry make the information infrastructure so important that some researchers consider it necessary to identify it as a unique form of capital goods in the classic production formula, which only differentiated between labor and capital. Maquilas also reflect the evolving spatial distribution of large corporations as a new international division of labor. Here the labor-intensive phases of the manufacturing process are done in peripheral countries. The important decisions are made in the headquarters located in the older industrialized countries, where high-tech activities and the service sector are consolidating.

Transborder Telecommunications for Technology Transfer

The Border as a Barrier to Transfer

Important in all cases is that maquila sites have excellent telecommunications with their parent companies, typically located in the United States. The telecommunication services that have been provided between the United States and Mexico, however, not only do not satisfy the needs of the industries already established on the border but have represented a major obstacle for the development and operation of new projects in the manufacturing and trade sectors. A related area of development is the extension of the so-called "economic corridor" that would extend from Austin to Saltillo. It is quite likely that as the Texan and the Mexican interests grow jointly in high-tech areas, cooperation with the state of Nuevo Leon will increase in importance. This process of cooperation can be enhanced substantially by the development of advanced telecommunications services. The corridor could also become the basis for cooperation in technology transfer in the form of cooperative education programs. The current exchange program and BITNET telecommunications link between the Monterrey Technological Institute (ITESM) and the University of Texas at San Antonio is an example of international links in this sector.

The telecommunications needs of the numerous maquiladora plants differ as widely as the nature of the operations themselves. These can range from the small family-run footwear plant where voice telephony is considered a cost for avoiding some trips, to second-generation plants owned by Fortune 500 companies where transborder data flow is an integral part of a worldwide data network and the craft paradigm is becoming relevant. Without advanced transborder telecommunications, there can be little, if any, transfer of advance production sharing. Several examples reflect different means for transborder telecommunications.

Telephone Networks: The Cases of Fisher-Price and General Motors

Fisher-Price, one of the world's largest toy manufacturers, is a subsidiary of Quaker Oats. It has an assembling plant in Matamoros, Tamaulipas, and has its respective offices just across the border in Brownsville, Texas. This plant has been doing real-time computing (with the mainframe located in Buffalo, New York) since December, 1984. This is one of the companies that has been using the services provided by TELMEX, the Mexican telephone company, and has two sets of data lines, one of them as a backup. Each data line requires two dedicated lines (2.4 kbps) because they are multiple-wire systems.

The terminal in Matamoros is connected with the two data lines to the Brownsville offices and from there to Buffalo, New York. This firm has a comprehensive master data system called Master Production System (MPS), which is comprised of three systems: the PLR, which deals with personnel matters, the MCS, which is the production-scheduling system; and the WYCS, which is an inventory system. The latter is a part of the corporation's Master Requirements Plan, which gives the New York offices the capability for immediate purchasing of supplies. The Brownsville offices can link directly to offices located elsewhere, such as the ones in Kentucky.

To date, the Matamoros terminal has had no technical problems with the telephone lines; the problems that have occurred have been caused by the computer equipment. However, transborder data communication needs have not been successful in other border points, especially Nuevo Laredo-Laredo and Ciudad Juárez-El Paso, where not even voice-communication services have been reliable in the past.

Another problem with the services provided by TELMEX is that the procedures for installation and repair of equipment were designed for the technical characteristics of analog systems. Under the old collective contract between the company and the union, each worker could only install or repair a limited number of units. The new contract not only gives management complete control over the schedules of the workers, but almost eliminates the role that the union used to have in the decision-making process for technological innovation.

General Motors (GM) is the largest user in the maquiladora industry, with 30 leased data lines in 27 plants. Telecommunications of this corporation are managed and operated by Electronic Data Systems (EDS), which has been a subsidiary of GM since 1984. This company still represents about 40% of the total work of the Services to the Maquiladora Industry Office of TELMEX, which was created in 1985 to attend to the demands of the industry in general and GM in particular. This office moved to Ciudad Juárez after a few months in Mexico City. The close working relation between EDS and TELMEX is starting to change because the GM subsidiary has plans to bypass the public network, using microwave links and coaxial cable; at the same time, it will become a competitor offering value-added services to other firms with maquiladora operations. This new provider uses an infrastructure that includes 633 private voice lines, 744 public telephone lines, 113 data lines, and 210 microwave channels.

Microwave Systems: The Case of Zenith

Zenith Electronics Corporation is one of the largest electronics firms in the United States. The principal research facilities are in Glenview, Illinois, with five manufacturing plants in the Chicago area and an assembly plant in Springfield, Missouri. Within the United States, Zenith has subsidiaries in Indiana (manufacturing television cabinets), Michigan (assembling microcomputer products), Texas (six locations serving as warehouses for materials moving between Mexico and the United States), and a variety of other locations throughout the United States established for display and sales.

Outside of the United States, Zenith owns a plant in Taiwan (manufacturing monochrome video displays and electronic components) and seven plants in Mexico. The plants in Mexico are located in Matamoros (manufacturing cathode ray tube electron guns and other electronic components), Reynosa (manufacturing and assembling small-screen television receivers, color television chassis, and module boards), and Chihuahua (one assembling cable products and one assembling power supplies). Two other plants are located in Ciudad Juárez.

Zenith is currently using two microwave systems for transborder communications, one in Reynosa-McAllen and another in El Paso-Ciudad Juárez. They are in the process of getting a permit to use a third one in Nogales-Tucson, but it is taking a long time to clear the paperwork with the Secretariat of Communications and Transportation (SCT). The network configuration permits the exchange of data or other information between any two U.S. or Mexico locations, but it requires going through the corporate headquarters. The Taiwan facilities are isolated from the network, and the exchange of information is done through the transportation of stored data. The firm is using the network to control the inventory, payroll, and production schedules of the plants in Mexico.

The main problem for Zenith and other companies related to the microwave systems is the long time it takes to clear the paperwork with the SCT. According to the Mexican Federal Law of Communications, microwave equipment cannot be owned or operated by private organizations, so they must obtain concessions from the SCT. This option was not designed for that purpose, however, so the potential users must hire the services of one of the independent consultants. These consultants, called *peritos* (there are between 500 and 600 nationwide), do all the technical paperwork, including soliciting letters from TELMEX specifying that the services provided do not satisfy the telecommunication needs of the firm. This has been a major problem, because TELMEX regarded microwave systems as bypasses competing for the same market and often delayed the paperwork. In the last few years, however, some bypasses have been permitted by the SCT without the letters by TELMEX. This new process will continue to occur more openly once the carrier is privatized.

Currently there are 31 bypasses, with 27 in El Paso-Juárez, three in McAllen-Reynosa, one in Brownsville-Matamoros, and another in Del Rio-Ciudad Acuña. Only five of the links are analog, and the rest are digital. The main providers of this equipment are two companies based in El Paso: Bordercomm, Inc., and Communications Diversified, Inc. This equipment has European standards because it has to be approved (or "homologated") by the SCT. It was expected that the new enhanced transborder services would discourage the installation of additional microwave systems for transborder data communication of private firms. Companies like Zenith, however, prefer to keep working with the current equipment and use the new services as a backup for redundancy.

Satellite: The Case of Westinghouse

The Mexican operation of Westinghouse, named *Sistemas Electrónicos Mexicanos,* is not the typical maquila operation for two reasons: It is not an assembly plant, but a computer software producer, and it is not located on the border, but in the city of Chihuahua. The location of maquila operations outside the border belt is an increasing trend, and cities such as Chihuahua, Monterrey, and Guadalajara are the sites of numerous plants. In the past, these plants faced more severe telecommunication problems than their border counterparts because they did not have the option of microwave systems and their telephone communications to border offices on the U.S. side were not billed as local calls.

Westinghouse was one of the first firms, along with Chrysler, to solve this communications problem with satellite transmissions. They have been using the services from INTELSAT's IBS for a Chihuahua-Baltimore link. This link is a part of the Westinghouse Private Network and is used to send software requirements and architectural designs from Baltimore. Software then is developed in Chihuahua by graduates and interns from Monterrey Tech (ITESM) and sent back to Baltimore for installation and operation, both inside and outside the firm. This link is essential to the business operation and is used extensively. Besides data transmission, it is also used for voice communications, facsimile, and audio and video transmission. Besides sharing with GM some of the characteristics of large users, Westinghouse also is going to become a provider of transborder communications.

As with microwave systems, it is expected that a new structure of TELMEX will provide similar services to firms located in the interior by using the Mexican Morelos Satellite System (SMS) and make the necessary arrangements to use other satellites when the downlink site in the United States is located outside the area of coverage of the system. South Texas and southern California, however, are well inside the footprint of the SMS.

Training and Shop Floor
Communication as Technology Transfer

The maquiladora industry of the last decade has been characterized by the co-existence of three types of operations: (1) those that involve easily acquired skills; (2) those that are labor intensive and require more highly skilled labor forces; and (3) those that can be described as flexible manufacturing or involve high technology, which in many positions need skilled workers. The last two particularly have grown in the last 10 years, and advanced telecommunication services not only facilitate the linkages between the engineering and managerial sectors on both sides of the border but can be used as a tool in enhancing the skills of the work force.

Training as Technology Transfer

Although the concept of technology transfer tends to evoke glamorous images involving scientists and consortia, it also can take the form of the adoption of discrete technologies or the training of the work force. The maquiladora industry is an example when firms import equipment on a temporary basis and do not establish strong links with the domestic industries, resulting in mere technology relocation. That is the case of the Ford-Mazda plant located in Hermosillo, which is one of the most sophisticated automotive plants in the world, having the highest number of robots anywhere. The first generation of shop-floor workers of this plant was trained for six months in Japan, and its satellite-communications facilities sometimes are used for training programs at different levels. The Westinghouse plant in Chihuahua sometimes will use the satellite for the same purpose, although the characteristics and level of its work force are very different. With many of the computer programmers coming from the Monterrey Tech (ITESM), it has donated equipment that includes state-of-the-art robots to the high-technology center of that institution.

Multinational firms have the need to customize the work force according to their needs because of the weaknesses of the education system of Mexico (attributable to the lack of resources of both the federal government and the students' families). This system has been characterized by: (a) a constant decrease in basic and middle education, where 336,000 have no access to the former, while the latter covers only 20% of the potential demand; (b) a decrease in subsidies for public universities in the last decade of 25%; (c) a reduction of funds for the National Council of Science and Technology (CONACYT), which forced the cancellation of 70 research projects, 600 new scholarships, and 35 projects for research centers in 1988; and (d) the small number of trade and technical schools compared to the number of liberal-arts departments in the universities (García Canclini & Safa, 1989). This also has caused some organizations, like the Bermudez Group of Ciudad Juárez, to participate actively in the design of the curriculum of trade and technical public schools (CBETIS and CONALEPs) in order to satisfy the needs of firms that might establish operations in 1 of the 11 industrial parks of the organization.

The use of communication networks is not restricted to switched networks, but also includes the more traditional broadcast networks. This is the case of the use of public television station in El Paso for distance education by the University of Texas at El Paso, with programming designed in cooperation with the maquiladora community. Another example is Santa Teresa, a master-planned community west of El Paso with more than 70,000 acres divided by the international border. Although this project is still in an early phase (currently working on road construction), there are plans to have an interactive cable system on both sides of the border that will be used as a complement to a magnet school in the areas of math and science.

The absence of indigenous R&D, aggravated by the budget and program cuts and the customization of technical education to satisfy the needs of the maquiladora industry, could be detrimental to the emergence of a domestic industry in Mexico that could be competitive in the global market.

One-Way Transfer

The new vital role of telecommunications has given more spatial flexibility to the multinational firms and has allowed them not only to do what GM calls "synchronous manufacturing," but to do it in a process where they have the ability to react rapidly to changes in either demand or technical innovations. This new role also has affected the patterns of communication within the firms. Synchronous manufacturing can increase horizontal and limited diagonal communication between units that are dispersed geographically, but at the same time it reinforces the unidirectionality of the flow, especially at shop-floor level.

The creativity model of technology transfer discussed in Chapter 2 can be applied in the maquiladora industry only to the R&D and engineering departments located in the laboratories and offices in the United States, but not to the lower levels of the organization. In the organizational structure of the firm there are two crucial interfaces: (1) between top management and R&D areas with the engineers and production supervisors at the Mexican plant; and (2) between the latter with the workers. The first interface tends to be a one-way flow but it allows for some form of reinvention at the production site, especially in flexible manufacturing operations that tend to have "quality circles." The second interface is completely unidirectional, however, where discrete procedures are imposed upon the workers through training or production-related instructions. This does not allow the workers any input on the conceptualization of either the production process or actual components, even when these have had impressive results when they have occurred. A clear example of this is a best-selling television set model of RCA that contained an entire circuit matrix designed by a group of line operators (Pena, 1983). Pena adds that 43.7% of the 215 maquiladora workers surveyed in Ciudad Juárez report to have contributed inventions or improved certain techniques at work, and that only 16% of them actually were remunerated. Some types of inventions include modification of production processes (11.5%), product or component design modification (8.7%), and modification of machinery or tools (17%).

The transfer of discrete production procedures without the input of workers may sometimes prevent a gain by the firm, but it also results in the creation of what Pena refers to as *technical surplus value*. The latter is created when labor adapts, repairs, or maintains tools and equipment—skilled activities for which workers are not trained or paid. In this sense, the firm can use telecommunications to provide the workers with the isolated skills that are more obviously related to the production process, and leave the rest for informal or self-training.

This pattern of unidirectionality is reinforced by the cultural differences between labor and management stemming from their demographic characteristics. Alain Lipietz (1986) describes the U.S.-Mexico border as "peripheral Fordism" because, among other things, the qualified employment positions remain largely external to Mexico. At the same time, Pena (1983) detects a recent trend of "Mexicanization" of production management because of its bilingualism and especially because it helps in establishing unity with line workers despite class differences.

Economic, Social and Political Ramifications

In our analysis of maquiladoras, we see that the process of technology transfer from a communication perspective has two main problems that affect the actors involved. One is technological, and the other deals with the dynamics of the process. The technological problem is the need for advanced telecommunications services that facilitate the new spatial strategies of multinational firms, and the latter problem consists in the unidirectionality of the flow. The combination of both problems can have a negative impact on individual firms, but it mainly inhibits the potential of Mexico in developing a competitive industry based on a more symmetrical sharing of production. This could be done only if there were more linkages between the indigenous industries and the maquiladoras in supplying and contracting, or if the characteristics of the Mexican middle management and operators employed in the maquiladora industry could have an impact in Mexican firms.

Clearly the easing of border restrictions on telecommunications, opening the way toward investment in advanced multinational production-sharing management, would boost the maquiladora industries. In both research and practical conversations about these businesses, the telecommunications barriers are heard about again and again. The alternative, however, of "opening" the advanced telecommunications systems for transborder growth—hence, production technology transfer—also has economic, social, and political ramifications.

There is a social consequence in Mexico if telecommunications advances are concentrated only on industrial applications. This ignores alternative use for social-oriented programs or substantive tasks, and it is critical because of the incongruity between the unbalanced use of the resources and the large segment of the population without basic telephone services. Underlying the new telecommunication policies of Mexico is the assumption that foreign investment will encourage development, and that the benefits will ultimately trickle down to the entire population. This assumption has proven wrong in the past in many Latin American countries, especially because most benefits return to the countries of origin of the large corporations, and in particular where the socioeconomic gap between different strata has increased. It is crucial for Mexico to decide how to use the telecommunications infrastructure to have the greatest impact possible in the sectors that need it the most, which means a need to promote rural development. Next, there are important long-range development questions regarding the future of advanced manufacturing methods in production-sharing environments. One is whether advanced manufacturing methods—as in the craft paradigm mentioned earlier—would indeed proliferate in maquiladoras if transborder telecommunications barriers were reduced.

As discussed by Wilson (in press), there is the continuing question of whether the low-wage basis for maquila growth is compatible with increasingly automated manufacturing techniques. They could, in the long range, be antagonistic. That is, as the percentage of labor costs are reduced through automation—assisted by production technology transfer—would that decrease at some point cancel the advantages of production sharing with Mexico? Furthermore, as Mexico recovers from its debt problems, or as workers become better trained in advanced manufacturing techniques, wages are likely to rise, thus contributing a further variable in canceling the maquila advantage. A negative scenario is that production-sharing telecommunications could make it easier for multinational corporations to move their production facilities once local wage advantages are lost. Maquila operations on the Mexican border might shift quietly through the emerging global network to, say, Haiti, or to Eastern Europe, where they could manufacture for both export and the new domestic markets.

On the other hand, despite the uncertain patterns of global industrialization, the strategic location and the *de facto* integration of the Mexican border to the U.S. economy makes its industrial future more secure than other visible alternatives at this time. Besides, Sassen (1988) points out that the flow of jobs is not always directed to countries with the least expensive labor, but to countries where labor can be controlled. Mexico is the best example of labor control through corporativist institutions, and the new work force of the maquiladora industry is not only demographically different (much younger and predominantly female), but according to Carrillo (1989) is more consensual (as opposed to conflictual). Others such as Gordon (1988) attribute the emergence of newly industrialized countries (NICs) to the diminished power of the multinationals since the mid-1960s, when corporate profitability declined dramatically. This decay forced these firms to relocate their operations to very selected export platforms where the economic advantages combined with stable institutional climates. This combination of variables proved to be a powerful predictor and included states willing to attract foreign investment by creating infrastructures with amortization times much longer than those of the plants and equipment of the firms.

The rapid implementation of new telecommunication services could put the U.S. firms on the border in a privileged position, but its planning must take into account the possible development of isolated networks that could later prevent their integration into Mexico's—as well as the

United States'—"networks of networks" because of the incompatibility of hardware and software. In other words, transfer of production sharing should be a collaborative matter, and not a one-way rich country-poor country "developmental" process.

Although the governments of the border states seem eager to initiate the process of "tertiarization" of the international projects, the Federal Government of Mexico has expressed in the General Agreement of Trade and Tariffs (GATT) negotiations its reluctance to mix the trade and service sectors. But at the same time, it has accepted that Mexico cannot afford to be left out of that process, as it was with the industrial revolution 150 years ago (Bravo-Aguilera, 1988). An increase of the integration of the Mexican border with the U.S. economy would also increase its asymmetric interdependence.

In the end, the Mexican state must decide whether the country will go with the flow of the ongoing restructuring of capitalism and continue with unpopular domestic policies, or be in control and generate programs that might decrease slightly the flow of international capital but aim at a more just society. The hard choices are economic, political, and social, which certainly are more profound than the important questions framed in terms of telecommunications support of production technology transfer.

References

Bravo-Aguilera, L. (1988). México frente a las negociaciones internacionales de servicios. *Comercio Exterior, 38*(1), 26-29.

Brown, F., & Domínguez, L. (1989). Nuevas tecnologías en la industria maquiladora de exportación. *Comercio Exterior, 39*(3), 215-223.

Carrillo, J. (1989). *Reestructuración en la industria automotriz en México: Políticas de ajuste e implicaciones laborales.* Tijuana: COLEF.

Davis, R. (1985). *Industria maquiladora y subsidiarias de co-inversión: Régimen jurídico y corporativo.* Mexico City: Cárdenas Editor y Distribuidor.

García Canclini, N., & Safa, P. (1989). *Políticas culturales y sociedad civil en México.* Mexico City: ENAH (mimeo).

Gordon, D. M. (1988, March-April). The global economy: New edifice or crumbling foundations? *New Left Review, 168,* 24-65.

Lipietz, A. (1986). New tendencies in the international division of labor: Regimes of accumulation and modes of regulation. In A. J. Scott & M. Storper (Eds.), *Production, work, territory: The geographical anatomy of industrial capitalism.* Boston: Allen & Unwin.

Maquiladoras: Production sharing Mexican-American style. (1988, February-March). *Expansion Management,* 26.

Mitchell, W. L., & Vargas, L. (1987). *Economic impact of the maquiladora industry in Juárez, Mexico on El Paso, Texas and other sections of the United States* [internal report]. Ciudad Juárez: Grupo Bermudez.

Notimex (1990, February 8). *México/Patentes* [wire service].

Pena, D. G. (1983). *The class politics of abstract labor: Organizational forms and industrial relations in the Mexican maquiladoras.* Doctoral dissertation, University of Texas at Austin.

Piore, M. J., & Sabel, C. F. (1984). *The second industrial divide: Possibilities for prosperity.* New York: Basic Books.

Poder Ejecutivo Federal. (1989). *Plan nacional de desarrollo 1989-1994.* Mexico City: SPP.

Sassen, S. (1989). *The mobility of labor and capital.* Cambridge: Cambridge University Press.

Sklair, L. (1989). *Assembling for development: The maquila industry in Mexico and the United States.* Boston: Unwin Hyman.

Wilson, P. A. (in press). The new maquiladoras: Flexible production in low wage regions. In K. Fatemi (Ed.), *Maquiladoras: Economic problem or solution?* New York: Praeger.

10

Japan: Tsukuba Science City

JAMES W. DEARING
EVERETT M. ROGERS

Tsukuba Science City represents one of the legends in model city planning to promote science development in our era. Approximately 100 small informal groups of researchers exist in Tsukuba to share research ideas and scientific results. The present analysis is based on a survey of small group leaders, and case studies of five of the groups. Results indicate that: (1) research study groups in Tsukuba Science City function to transfer technology among researchers, but that the value of the knowledge transferred varies according to whether the groups have industrial members or not; (2) small group communication does, in some cases, lead to collaborative research projects among group participants; (3) the function of research study groups has changed since their inception, perhaps becoming less important over time; and (4) research study groups are most effective when they bridge relational gaps between academic, government, and industry researchers. James W. Dearing is Assistant Professor in the Department of Communication, Michigan State University. Everett M. Rogers, well-known as coauthor with Judith Larsen of Silicon

Valley Fever, *is Walter H. Annenberg Professor of Communica-*
tions at the Annenberg School for Communication at the Univer-
sity of Southern California.

A Plan for Technology Transfer

What are the most effective means to transfer technology among
researchers in a single geographic area? This question faced Japanese
science policymakers in the mid-1970s as they contemplated how best
to stimulate the development of interpersonal communication networks
among scientists and engineers in Tsukuba Science City, Japan's large-
scale science experiment. The economic and political obstacles of
establishing a city of science could be overcome by government action,
but the social obstacles of creating a community in which researchers
would live and be optimally creative were not solved so readily by
legislation or money. An atmosphere of intellectual and social communication had to be cultivated in which researchers could pursue their
work with enthusiasm. Work and enthusiasm, together, can lead scientists to bright ideas (Weber, 1958).

In 1979, thousands of researchers who had no previous contact with
each other moved from the anonymity of Tokyo and began to work in
the quiet new science city. Many of these researchers were about 30
years of age and felt that their new surroundings might free them from
the restrictive research environments they had known in Tokyo. These
young researchers felt that government research in Tokyo was dominated by cumbersome administrations and senior researchers and was
of little relevance to Japan's practical problems. Whereas the contributions of these young researchers might have been minimal in Tokyo,
relocation to Tsukuba Science City offered them the opportunity to
prove themselves to the scientific establishment in the capital.

Top-ranking Japanese science administrators tried several means of
encouraging more communication among their Tsukuba researchers,
but researcher participation and enthusiasm were low. Scientists prize
their personal autonomy from bureaucratic authority (Merton, 1968).
Moreover, research institute administrators were more accustomed to
competition than cooperation, so their efforts at communication were
halfhearted (Kawamoto, 1987).

The new Tsukuba researchers recognized that the organizational and administrative barriers that had segregated their research institutes in Tokyo had not yet developed in Tsukuba. Everything was new. The researchers sensed the opportunity to meet colleagues in other Tsukuba research institutes in order to share ideas. By 1982, about 80 informal small groups of researchers were meeting regularly. Small groups can be effective means of technology transfer, because it is through inter-personal communication that people transfer technology (Larsen, 1988).

This chapter investigates the status of research study groups in Tsukuba, and the extent to which these groups of researchers—which now number about 100—have been successful in transferring technology among their participants in the sense of the communication of knowledge among the participants of a social system.

Tsukuba Science City Today

In May, 1989, Tsukuba Science City had a daytime population of 162,189 residents, 47 public research institutes, and 177 private firms operating or committed to begin operation in Tsukuba that either maintained R&D labs, manufactured high-tech products, or served as research surveillance offices. Total employment by public and private research institutes, including the two universities in Tsukuba, is about 17,500. More than 9,000 publicly and privately employed researchers work in Tsukuba, 7,000 of whom are government researchers. Of the 7,000 government researchers, approximately 2,600 hold doctoral degrees. In this small city, 45% of Japan's national researchers work in 33% of Japan's national research facilities, supported by 48% of the nation's research and development budget.

From 1979 to 1990, research activity in Tsukuba has been dominated by the government researchers who work at the 47 national research institutes in the science city. These institutes, which include two universities, are located in the center of the city. Around this core is a rapidly growing number of private R&D laboratories. Whereas hiring and funding levels at the national labs have remained about constant from 1980 to 1990, hiring and funding levels are rising rapidly at the private labs. By the year 2000, research activity in Tsukuba Science City is likely to be dominated (in terms of the number of labs, the number of employees, and research budgets) by private industry (Dearing, 1989).

The Japanese national government, especially through its Science and Technology Agency (STA) and the Ministry of Education, encourages technology transfer among publicly and privately employed Tsukuba researchers. An STA program in which young, privately employed researchers take a leave of absence from their laboratories and spend from one to two years as research assistants to senior researchers in the national labs appears to be successful. For both the researcher and the organizations involved, there are distinct advantages to collaboration. Privately employed researchers tend to be young graduates without much experience or strong theoretical backgrounds. They work with other young researchers, and their projects are limited by the projected commercial feasibility of the research. By working as assistants in university departments and national research institutes, private-company researchers gain experience and guidance from senior researchers. Their firms benefit by tapping the knowledge of senior researchers, which may then be reflected in more innovative work by their employees.

Both university and national institute researchers tend to be less informed about the latest instrumentation and apparatus that private company researchers take for granted. The turnover time for lab equipment is more rapid in private companies (Science and Technology Agency, 1988). University and national research institute researchers have less funding with which to conduct research; their main dilemma, however, is a staffing shortage. Wages are comparatively low, so it is difficult to attract the brightest young researchers. Both university and national institute researchers use university graduate students as research assistants, but young privately employed researchers tend to be brighter, have some experience, and have their salaries paid by the participating companies. The universities and national institutes benefit by "retooling" the knowledge of the faculties and employees about state-of-the-art equipment and technologies, and they save money by not paying salaries. It is a highly symbiotic technology transfer system.

Tsukuba City Survey and Case Studies

Method

In order to investigate the status of research study groups in Tsukuba, and the extent to which these groups of researchers have been successful in transferring technology among their participants, we used both (1) survey questionnaire and (2) case study methodologies.

A three-page questionnaire was mailed to researchers listed as the organizers of research study groups in Tsukuba. Seventy-six contact persons were identified from a list compiled by the Tsukuba Center for Institutes in Spring, 1988. This list was created by assigning a secretary to telephone each of the national research institutes in Tsukuba and to ask administrative persons in charge of research whether they were aware of any of their researchers meeting outside of work in research groups. Thus, this list of 76 was determined by the information (or lack thereof) that institute administrators had about their researchers' activities.

Three strategies for supplementing this list were followed. First, a Tsukuba resident experienced in academic research was hired to go to each of the institutes and collect information from the notices posted on bulletin boards about research group meetings. Secondly, three messages (two in Japanese and one in English) were placed on the "bulletin board" of a new electronic data base linking researchers in Tsukuba. Members of research groups were encouraged to either telephone the author of the message or send an electronic message to a certain user number. Third, back issues of *Science Communication*—a locally distributed poster printed by the Tsukuba Center for Institutes— were examined for possible contact names of study group organizers. *Science Communication* is a popular means by which study groups advertise their meetings.

These supplementary strategies resulted in the identification of a possible 65 additional research groups. A questionnaire was drafted in English and translated into Japanese; questions in Japanese were then tested for the original English meaning, and adjusted. One scientist and one science administrator in Tsukuba reviewed the Japanese-language questionnaire and made criticisms and suggestions for revisions. Several questions were added and coding categories were revised. The complete mailed packet consisted of: (1) a Japanese-language cover letter written by Tetsuzo Kawamoto, a widely known and respected science administrator in Tsukuba Science City; (2) an English-language cover letter and explanation written by one of the present authors; (3) a photocopy of a newspaper article about the present research investigation, which had been published locally in Tsukuba two weeks before the packets were mailed; (4) the three-page questionnaire; and (5) a return-addressed stamped envelope. The packet envelopes and the return-addressed envelopes both were printed with the name of the Tsukuba Research Consortium, a private organization in Tsukuba that is well-known for science communication activities. One hundred and forty-three packets were mailed.

Within one week, 30 completed questionnaires had been received. After three weeks, 65 completed questionnaires had been received. Follow-up telephone calls brought the total number of completed questionnaires to 71. Judging from some mailed and telephoned responses, perhaps as many as 40 of the 143 total mailed packets went to the organizers of meetings that were not topic-specific informal study groups of researchers who meet regularly. The population of such regularly meeting groups is not known exactly, but given a reasonable estimate of about 100 our survey response rate was about 70%.

Respondents were asked to identify the institutional affiliations of the members of their research groups and the number of researchers from each institution who were group members, as well as to respond to several other closed- and open-ended questions.

Our case studies of five informal groups of researchers were based on personal interviews with group organizers, leaders, participants, and nonparticipants. Researchers averaged about 40 to 45 years of age. They were identified by (1) elite researchers in Tsukuba; (2) a list compiled by the Tsukuba Center for Institutes; (3) each other; and (4) through publications. Personal interviews with these group leaders averaged one and one-half hours. Selection was not random, because very specific information was sought from individuals with unique knowledge. Our distribution of interviews showed a bias toward government employed researchers because the groups had more participants from the government research institutes than from universities or from private firms.

Measuring the success of the communication initiatives of study was done by listening closely to and interpreting how interviewees appraised each communication initiative. The main determinants of the degree to which initiatives were successful were the perceptions of the interviewees and our observation of the groups and their members.

Our individually tailored questionnaires were open-ended, and many follow-up questions were asked. Almost all of the interviews were conducted in Japanese and English, with the assistance of interpreters. Some interviewees (particularly younger government researchers) had a good command of English because of their graduate studies in the United States. A number of interviewees who felt at ease speaking in

English did so, resorting occasionally to Japanese by consulting with the interpreter or with the investigator. For interviewees who spoke adequate English, Japanese interpreters were often brought along as a courtesy, so that they could speak in either Japanese or English. To make sure that the content of our questions was understood, questions were asked most often in Japanese through interpreters.

Survey Results

The present data are based on written responses from the leaders of the 71 study groups. Some of the informal study groups in Tsukuba Science City have as few as seven members, although most study groups have about 30 members. The number of active participants at any particular meeting is about half of the group's total membership; topics range from catalysis to wind tunnels to viruses.

Researchers who are members of Tsukuba Science City study groups overwhelmingly are middle-aged. Only one study group's membership averaged between 21 and 29 years of age, and only one study group's membership averaged over 50 years of age; 85% of the study groups had memberships that averaged between 30 and 44 years of age; 66% of the study groups met less frequently than once per month, while 34% met once a month or more frequently.

Answers to the question, "How formal is your study group?" show that most of the study groups are rather serious gatherings. Meetings center around the presentation of research results or of research in progress. Nearly 22% of the study groups alternated the style of their meetings, mostly between formal lectures and less-organized free discussion.

Many of the study groups had members employed at different institutes who were involved in collaborative research. To the question, "Have there been any joint research projects or joint research activities between members of your study group who are from different institutes?" nearly 32% of the respondents replied that their groups had members who had participated together in joint research. An additional 23% thought that this was the case, but were not sure; 7% of the study groups had members who were planning collaborative research.

Of crucial importance to the issue of collaborative research is whether the idea for the collaboration originated from the study groups or not. Of the 30 study groups in which collaborative research had occurred, was in progress, or was being planned, half of the respondents thought that the projects had been or were likely to have been inspired by bringing the collaborators together as a result of participation in a study group.

The context of communication between researchers in Tsukuba Science City has changed greatly over the past 10 years. There is much more scientific communication, and there are many signs of maturity in Tsukuba Science City itself. Thus, we were interested in determining the degree of change of the informal study groups. We asked, "Has your research group changed over time (especially in the ways that members make presentations or discuss research in meetings)?" Almost 52% of the study groups had not changed much over time, and 80% either had not changed much or only changed a little. In open-ended replies, slightly more respondents implied that their groups were changing for the worse rather than the better, with comments such as "Everybody is getting too busy to meet," or "There is a decreasing number of members," or "It's getting formal and ordinary, and discussion is less hot." Positive responses included "We have parties now," and "We are paying attention to content more, rather than just introducing the field of study."

We also were interested in why researchers attend study group meetings. We asked: "What do you think should be the main function(s) of your study group?" Researchers mainly see study groups as opportunities to get new ideas, to talk with colleagues about research problems, and to stimulate their own research. Open-ended responses frequently cited the possibilities of exchanging information and increasing research collaboration.

When asked, "How successful do you think that your study group has been?" respondents overwhelmingly considered their study groups to be either fairly successful (79%) or very successful (17%). Only 3% responded that their groups were not very successful.

When asked the open-ended question, "What do you think is the ideal type of new member for your group?" about one third of respondents wrote that energetic people were most important, as well as those researchers who were inquisitive and had different viewpoints. Only a few respondents specified that new members should be young.

The first study group was formed in 1979. By 1982, the number of study groups already had reached about 80. Why did the number of study groups grow so rapidly between 1979 and 1982, and then slow considerably after 1982? The answer is not clear; perhaps this number more or less covers the intellectual breadth of research topics in the Tsukuba area.

A more plausible explanation is that at the time when most of these study groups were formed (1981-1982), the majority of the research population had just moved to the new science city. Thirty-six national institutes were open and at least partially operating by 1980. The organizers of the various study groups invariably were young researchers in their 30s who wanted to give the new science city a distinct identity apart from Tokyo. For them, creating and leading a study group that purposely crossed organizational boundaries and invited participation from university professors and privately employed researchers was a new and somewhat radical concept; they were innovators in adopting a new type of scientific communication. The organizers (many groups had several organizers) identified closely with their groups, and they took pride in their accomplishments of communicating in a new way. Some of the groups met in the Tsukuba Center for Institutes—a neutral meeting place, because members were from different institutes and some were under the jurisdiction of different ministries.

It is conceivable that a large degree of the satisfaction and rewards derived from participating in research study groups results from creating the groups. To start a new research study group now in Tsukuba certainly would not be a radical concept. Researchers who would try to do so might be rebuked socially by their senior researchers, who would prefer that the younger researchers join the appropriate established study groups. "Everybody likes to have their own club. I think that there are too many study groups," said Takeshi Fujita, a study group member and the director of an academic liaison program for Eizai Company in Tsukuba.

Case Study Results

We conducted case studies of five small informal groups of researchers who met regularly in Tsukuba. These five groups were: (1) the Tsukuba Applied Geoscience Society; (2) the Genetics and Bioengineering Study Group, (3) the Thermo-Fluid Dynamics Seminar, (4) the Heat Transfer Colloquium, and (5) the Tsukuba Chemical Science Club.

The Tsukuba Applied Geoscience Society (TAGS) is an example of how an informal, researcher-controlled study group can bring together researchers from many organizations. TAGS has members who are researchers in the National Research Center for Disaster Prevention,

Geographical Survey Institute, University of Tsukuba, University of Chiba, Public Works Research Institute, Building Research Institute, National Institute for Environmental Studies, Mechanical Engineering Laboratory, Geological Survey of Japan, National Institute of Agricultural Science, National Research Institute of Agricultural Engineering, and the Forestry and Forest Product Research Institute. These researchers share an interest in applied geoscience, which combines studies of geography, vegetation, surface soil science, geology, and hydrology in a focus on the prevention of natural disasters. In 1989, members of TAGS completed a collaborative investigation of slope failures that was funded specially by the Japanese Science and Technology Agency.

Active members of TAGS, nearly all of whom have been participants for 8 to 10 years, cite several nagging problems of their study group: (1) the group has lost its interdisciplinary nature; (2) meeting attendance has decreased; (3) there is a lack of interest in the group by young researchers; (4) meetings have become repetitive and sometimes boring; and (5) socializing after the monthly meetings has decreased. Annual field trips to areas of geologic interest still attract many members, but the monthly meetings do not. Why is TAGS faced with these problems, how serious are they, and how might they be remedied (if, indeed, they should be)?

TAGS initially attracted many meeting participants of diverse specialties partly because it was the only geoscience group meeting in Tsukuba. Gradually, however, competition arose in the form of other, more exactly focused geoscience study groups. For example, a group specializing in tectonics (the movement of earth's plates) began. Another group formed to discuss basic geology. Thus members who had made TAGS such an interdisciplinary group began to attend other meetings that they perceived as being more relevant to their specialized research work. This winnowing of its original eclectic nature left TAGS increasingly narrow in its topics of discussion, because only those researchers most interested in applied geoscience continued to attend meetings regularly. Thus, the group has lost membership at its disciplinary extremes, no doubt making the meetings increasingly predictable and similar—and, hence, less interesting to many.

Both the Heat Transfer Colloquium and the Tsukuba Chemical Science Club are examples of thriving study groups. Each group has strong ties to private researchers and engineering and chemistry professors, as well as many members from the government research institutes in

Tsukuba. Organizers of the groups consider their membership diversity to be a major reason why their members are enthusiastic about the study groups. Collegiate friendships often serve as the relational basis for study group organizers when they search for industry and academic researchers to join their group.

In both the Heat Transfer Colloquium and the Tsukuba Chemical Science Club, the placement of graduate students with private firms is a main outcome of the interpersonal relationship between professors and industrial researchers. Privately employed researchers join the groups partly to meet professors and secure an inside track on the availability of the best graduates; professors join the groups partly to meet employers and to place their students with the best companies.

The Genetics and Bioengineering Study Group has halted virtually all its activities because the research environment in Tsukuba has changed, making the existence of the group less necessary. This group's rise and fall seems to have paralleled the excitement of developments in the field of genetics in the early 1980s, as well as to have corresponded with the newness and subsequent adolescence of the science city. Clearly there are times within disciplines and at points during research-paradigm formation when informal study groups play important roles in diffusing scientific knowledge. Just as clearly, informal study groups can outlive their content contributions to a set of researchers. The necessity for the group, which was very strong at first, eroded when the Ministry of Agriculture, Forestry, and Fisheries opened the National Institute for Agrobiological Resources in December, 1983. This center brought together most of the researchers from various institutes who had been attending the Genetics and Bioengineering Study Group's monthly meetings. Also, private genetics research blossomed worldwide as the profit potential of biotechnology became apparent. Trade journals and academic journals proliferated, making knowledge about genetics easier to get.

As more researchers have moved to the science city, informal study groups have tended to become less interdisciplinary and more intradisciplinary. It is easier to establish and interest researchers in initiatives that are intradisciplinary. Researchers recognize the value of interdisciplinary initiatives, but they tend not to participate in them.

Leaders were important in the establishment and operation of the study groups we analyzed. Each group was formed and guided by a strong-willed, energetic researcher; several of these were known as excellent researchers by their peers.

The value of the technology transferred among group members differs according to whether groups included researchers from private companies. In study groups without industrial members, or in which the topics of discussion are perceived to have little industrial application, the value of the knowledge discussed is relatively high. In study groups with industrial members, or in which the topics of discussion are perceived to have industrial application, the knowledge discussed is less valuable. For example, both the Heat Transfer Colloquium and the Tsukuba Chemical Science Club have many members who are industrial researchers. Because of the competing interests among company researchers, the degree of specificity of the engineering and, especially, chemical knowledge discussed in the study groups is slight. Membership in study groups bridges the gap between (1) very formal communication, such as the written reports that academic and government researchers prepare based on company-funded research grants to Japanese universities and government research institutes; and (2) very personal communication, such as paid, private consultations among professors and government researchers on the one hand, and company researchers on the other hand. To become an intimate member of the more valuable academic-government-industry personal networks requires a means of personal entry that the very formal institutional relationships do not provide. Research study groups, which bring people together albeit in a somewhat formal atmosphere, provide the social opportunity for entry into the valuable, very personal communication among researchers.

Lessons About Small-Group Technology Transfer

In retrospect, what is the status of research study groups in Tsukuba Science City, and how effective are these groups at transferring technology among researchers in Tsukuba Science City?

Approximately 100 small informal groups of researchers existed in Tsukuba to share research ideas and results. Research study groups in Tsukuba Science City function to transfer technology among researchers, but the value of the knowledge transferred varies according to whether the groups have active members who each are interested in key phenomena, but who come from divergent organizations and have different uses for the phenomena of study. Study groups with strong representations of industrial researchers appear to be the most intellectually active groups. Meetings tend to begin as serious academic discussions, and may be followed by informal drinking-and-eating parties afterwards.

More than 60% of study group organizers replied that their members either had conducted collaborative research among themselves, probably had done so, or were planning to do so. Half of these survey respondents believed that collaborative research projects resulted from discussions in the study group.

Survey and case study results indicate that the function of research study groups in Tsukuba has changed since their inception. Most of the groups, which can be categorized as minor rebellions against the established organization of research that prevailed in Tokyo during the late 1970s, have themselves become institutionalized. As a result of a changing research environment in Tsukuba and its scientific disciplines, not all of the groups that once were effective in technology transfer are still so effective. The institutionalization of study groups seems to reflect the intellectual conservatism that characterizes aging group leaders. This institutionalization can serve as a disincentive for young researchers to join the group.

In form and function, study groups respond to the scientific disciplines they address. As disciplines change, so do the study groups. If a study group fails to change with science, the group eventually is abandoned by its membership. It is not very difficult to organize an exciting study group if the topic is exciting and in the process of reformation on a worldwide basis. In such cases, knowledge is at a premium, so study group meetings are well attended.

We found that study groups are effective at bridging relational gaps between academic, government, and industry researchers, and at establishing the personal networks that may then be used for very specific goal-oriented technology transfer. The technology that may be transferred in small study groups is, in general, of lesser intellectual value than is the higher-value technology that may be transferred through personal networks.

For technology transfer, study groups are most important in that they give birth to interpersonal networks. Researchers meet and become familiar enough to trust one another through study group participation. The technology they exchange is of relatively low value, but common or complementary research interests—as well as social interests—may then lead to a strengthening of friendships. These interpersonal relations then serve as the mechanisms for transferring relatively high-value technology among researchers.

The present results may be usefully applied to the organization of informal research-based groups in communities other than Tsukuba Science City. For example, informal research-based groups should address topics of keen interest to researchers and should be based on knowledge that is not published yet. To attract members, group meetings should be perceived as valuable opportunities to learn about cutting-edge research. Also, the quality of group discussions and the extent of person-to-person information exchanges borne from interpersonal contact within the group can be heightened by attracting members with different goals and backgrounds who share an interest in the topic around which the group is organized. In particular, we found that study groups centered around research fronts that are the products of two or more fields of study often are characterized by high degrees of intellectual ambiguity and memberships of diverse specialists, all sharing an interest in a new topic.

Study groups also should be identified with—and at least initially led by—well-respected researchers whose presence and contacts will attract members. Organization of the meetings and group publications typically are handled by younger members. Young, diverse researchers should be brought into a group and given organizational responsibilities on a regular basis, so that the group will not become identified overly with a small clique of people. Study groups are perceived as interesting by researchers when the groups are change oriented, not only in content but also in structure. Organizational responsibilities and, to an extent, leadership roles should be transient and accessible to outsiders. The institutionalization of informal study groups is often an indication that the group increasingly serves the function of reinforcing the social status of the researchers involved, and decreasingly serves the function of transferring knowledge and technology.

References

Dearing, J. W. (1989). *Communication among researchers in a science city: Tsukuba, Japan.* Doctoral dissertation, University of Southern California, Los Angeles.
Kawamoto, T. (1987). *Speaking frankly on research communication in science and technology.* Unpublished paper, Tsukuba Research Consortium.

Larsen, J. K. (1988, December). *Lessons learned about technology transfer.* Paper presented at the Symposium on Science Communication: Health and Environmental Research, Los Angeles.

Merton, R. K. (1968). *Social theory and social structure.* New York: Free Press.

Science and Technology Agency. (1987). *White paper on science and technology 1987: Toward the internationalization of Japan's science and technology (summary).* Tokyo: Foreign Press Center.

Weber, M. (1958). Science as a vocation. In H. H. Gerth & C. W. Mills (Eds.), *From Max Weber: Essays in sociology.* New York: Oxford University Press.

11

Italy: Tecnopolis Novus Ortus and the EEC

UMBERTO BOZZO
DAVID V. GIBSON

Tecnopolis Novus Ortus is located near Valenzano, a small town in the Apulian region of southern Italy. The Italian technopolis is modeled after the Japanese effort to move high-tech research and development out of congested city areas and to spur regional economic growth and a high quality of life. In this chapter the authors emphasize regional high-tech development in the European Economic Community (EEC) through advanced telecommunication services as exemplified in Tecnopolis Novus Ortus. Umberto Bozzo is the General Director of Tecnopolis Novus Ortus. David V. Gibson's institutional affiliations are described in the introductory chapter of this book.

The Global Market

In the final days of the 20th century, the unrestricted flow of information and ideas across borders is essential to the well-being of the planet and its people. Cooperation in technology among industrialized nations is no longer simply a necessity, it is fast becoming a critical testing ground to determine how far friendship or confrontation is to develop between countries. Technology and its transfer has become, quite simply, a political issue. . . . The exchange of scientific and technological information is . . . a requirement imposed by the quality of our lives and those of future generations if not by our very survival on this planet. (Petrignani, 1990)

The global market is characterized by a number of features having world relevance such as patents, international standards and regulations for products and goods, the flow of information through telecommunication networks, and the necessity of having international markets to increase domestic income (Petrella, 1989). Many and important are the consequences of these trends, which will have implications on R&D, technology transfer, universities, public administrators, and business.

The development of the global market in economics and technology is strongly connected to (1) the integration of technologies such as computers and telecommunications, new materials, and microelectronics; (2) the integration of sectors such as agriculture and chemistry, telematics and financial services; (3) increasing costs for R&D; (4) reduction of product life cycles; (5) the uncertainty of markets; and (6) the shortage of specialized manpower. Given these trends enterprises look for cooperation and joint ventures with increased frequency, and this search is not limited by industry sector or by geography.

Acquisitions, mergers, technology, and commercial cooperative agreements are modifying the structure of many firms and markets throughout the world. The creation of a single European market by 1992 (Friberg, 1989; Magee, 1989; Snowdon, 1988) is in line with the present trend toward a single global market. Described in a 1985 EEC (European Economic Community) commission white paper, the internal market program consists of 285 legislative directives intended to eliminate barriers to the free movement of goods, people, and capital among the 12 EEC member states. Internal market directives will reach into every aspect of commercial activities, from the elimination of border controls and duplicate customs documents to the harmonization of product standards and guidelines on company mergers. A single internal marketplace with freedom of movement for goods, services, people, and capital will result in a $4 trillion market of 320 million people.

European officials believe that the integration of the EEC will increase economic growth and employment, and lead to greater consumption and imports. A study by the EEC Commission predicts that the removal of existing barriers may result in a 5% increase in EEC gross domestic product—more than $260 billion—through economies of scale and greater economic efficiency. Although European industry should receive the most direct benefits of a single European market, the EEC also will change substantially the way foreign companies do business in Europe. The ability to compete in a continental-scale market and to avoid duplication of administrative procedures, production, marketing, and distribution systems will offer great challenges and advantages.

An Italian Experiment: Tecnopolis Novus Ortus

In 1969, the University of Bari promoted the seeds of the technopolis concept in southern Italy, with the cooperation of Italian enterprises and local and national government. The creation of CSATA (Centro Studi Applicazoni in Tecnologie Avanzate) was the first step in forming Tecnopolis Novus Ortus. CSATA initially was located on the campus of Bari University and was affiliated closely with the computer science and physics departments. In 1984, CSATA moved into its present facility at Tecnopolis Novus Ortus, about 15 kilometers from the university and the city of Bari. This emerging technopolis is modeled after the Japanese technopolis concept, which stresses locating high-tech research facilities out of congested urban areas (Smilor, Kozmetsky, & Gibson, 1988; Tatsuno, 1986).

In 1988, CSATA changed its name to Tecnopolis CSATA Novus Ortus (TCNO) in order to stress its link to the technopolis concept. TCNO is situated near Valenzano, a small town in the Apulian region that traditionally has been an important meeting place for the different peoples and cultures of the Mediterranean basin. The region has about 4 million inhabitants, two universities (Bari and Lecce) with a combined student population of 63,000, and 16 public institutions and research centers. In 1951 there were 753,000 Apulians involved in agriculture and 494,000 involved in the industrial and service sectors. By 1985, the agricultural sector had decreased to 319,000 persons and the industrial and services sector had grown to 985,000.

TCNO offers businessmen throughout the Mediterranean the opportunity to take advantage of the technopolis strategy of collaboration between advanced and developing countries. For example, Tecnopolis Novus Ortus is the seat of the Community of Mediterranean Universities (CUM), which brings together more than 80 universities in projects aimed at scientific and education cooperation. It has been instrumental in setting up links with large private concerns operating in high technology, such as IBM Italia, Olivetti, Italtel, Sip, Fiat, CCC (Computer Curriculum Corp.), Datanet, Dataconsyst, RLG Inc. (Research Library Group), and Sesa Italia.

A major goal of Tecnopolis Novus Ortus is to spur high-tech economic development in the southern part of the country. Traditionally, Italy's economic growth has been in the north in such areas as Milan and Turin. Political, industrial, and academic leaders in general have long viewed the south as coming in a poor second in terms of education and industrial strength. Most of the large Italian companies have their headquarters in the north, and the most prestigious Italian university is in Milan.

The developmental gap between the southern and northern industrial areas is demonstrated in per capita income, which averages about $100 for the north and $68 for the south. The northern industrial regions, however, are experiencing problems that inhibit continued industrial development. First, there is limited room to expand industrial facilities. Second, the demand for skilled workers exceeds the available supply. For example, the demand for electrical engineers is double the university output.

Tecnopolis Novus Ortus was designed in the 1970s with the goal of a campus-like facility. The strategy was a very pragmatic, step-by-step approach. There was not a lot of formal planning. The idea was to create, on a small scale, all of the functions and amenities thought to be important to a larger technopolis. These facilities included a state-of-the-art conference room, a well-appointed restaurant, recreational facilities, ample office and lab space, an international and well-funded library, data banks, and educational facilities. The initial facility was 11,000 square meters on 4 hectares; there are currently six attached buildings. As of 1988, the consortium running the park employed 200 professional and support staff. By 1990, TCNO is expected to grow to 250 employees. In a sense, Tecnopolis Novus Ortus is a two-tiered organization: one of professionals, and one of clerical support.

In many ways, Tecnopolis Novus Ortus exhibits an unusual atmosphere for a southern Italian company. It seems to visitors more like a Silicon Valley high-tech firm than an indigenous Italian operation; perhaps the preferred culture of high-tech firms or think tanks transcends national boundaries. This technopolis seems like an oasis in the dry Bari landscape in that TCNO has such amenities as fountains and landscaped gardens, as well as tennis and handball courts. The one large on-site restaurant, for all employees, provides an ample selection of quality food. For after-meal relaxation and informal get-togethers there is an espresso bar with pastry, ice cream, and other snacks. Such a high-quality working environment does not exist in most Italian firms.

Employee ties to Tecnopolis Novus Ortus are strong. There are many instances of employees who have moved north and later wanted to return to Tecnopolis—in part because it is an exciting and enjoyable place to work with its unique mission and culture, in part because of Bari's climate and life-style, and in part because of strong ties to family in the Bari region.

Cohesion Through Telecommunications

As information becomes the world's most traded commodity, it is emerging rapidly as a key production factor. The consequences of this trend on the strategies and actions for regional economic development have been studied and referred to in the necessity for revising traditional schemes and concepts (Goddard & Gillespie, 1989; Huber, 1987; Smilor, Gibson, & Kozmetsky, 1989).

According to the most up-to-date development models, the basic conditions for the creation and growth of an information economy in a particular region can be found in the availability of suitable telecommunication infrastructures and a new range of goods that are referred to as knowledge, skill bases, and services in the information field. Telecommunications and service structures are the key factors for regional development, as they are able to attract and to benefit the location of business activities as well as linkages with outer markets.

If, on one hand, it is recognized that telecommunication networks are a strategic pulling factor for development, on the other hand several aspects and issues (ranging from technical to service and regulatory) need to be considered. In particular, experimentation of new application patterns and services, the stimulation and growth of demand for advanced networks and information services, the training of users, and the creation of a qualified skill base are all interrelated factors of high importance.

Regional Development Through Advanced Communication Services: STAR

It is recognized in the EEC that there exists a growing need for specialized information for effective decisions in business, political, scientific, and technical activities, as well as for professional, social, and economic choices. This need requires the adoption of advanced technology and the ability to determine the development of a specialized information market, which is to be connected and integrated to related European initiatives and markets.

In autumn 1986, the European Commission launched the STAR (Special Telecommunications Action for Regional Development) program to help less developed regions implement the necessary communication infrastructures. STAR is becoming one of the main tools used to bridge the technological gap between the member states and to improve the cohesion of the European community. Between 1500 and 1700 million ECU (European Currency for the EEC) will be invested over the 1987-1991 period within the framework of the STAR program to procure basic equipment for advanced telecommunications services and to promote applications of these services in certain regions of Spain, France, Greece, Italy, Portugal, Great Britain, and Ireland.

The STAR program strategy at Tecnopolis Novus Ortus Research Park started as an experimental project (Bozzo, Gibson, Sabatelli, & Smilor, 1990). As a preeminent feature of the project, both technology transfer experts and specialized equipment are being made available to selected users at their own premises. Demo training and consulting activities are performed on-site to develop the users' awareness of the strategic value of information and its organizational, educational, economic, and technical implications (Figure 11.1).

One main objective of STAR is to stimulate the demand for specialized information, especially in those sectors for which the access of such information is an essential strategic factor of economic development and decision making. The program has significant promotional

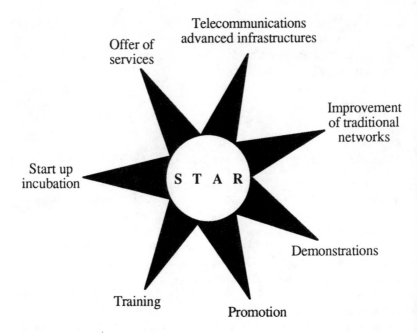

Figure 11.1. The STAR Program to Assist Less Developed EEC Regions
Implement the Necessary Communication Infrastructure

features because one of its main concerns is to develop and to diffuse
awareness about the strategic function of information resources in
such user segments as small- to medium-size firms, local government,
and administrative, economic (credit and commerce), and technical
institutions. This is accomplished by both demonstration activities
carried out at selected user sites and larger-scale simulation and promotion initiatives.

A second main objective of STAR is to prepare the conditions for,
organize, and make available adequate, specialized information services for the benefit of several user classes. On the supply side, the
project stresses the aspects of:

- setting up and providing information services on the basis of available
  technologies and capabilities; and

- showing the possibilities for organizing, producing, and handling information, as well as for improving and managing information services in order to set up a favorable environment for the establishment and growth of local entrepreneurial initiatives dedicated to the marketing of information resources.

A Multifunctional System for
Telecommunications and Telematic Services: NETT

In carrying out the STAR project, Tecnopolis Novus Ortus has exploited and enhanced its professional and equipment capabilities by working with an advanced multifunctional system for telecommunications and telematic services (NETT). NETT is integrated with the whole set of technological resources existing in the technopolis, and it is made available to local communities. NETT is also integrated with the most qualified technology people in Italy and abroad through existing transmission networks. This objective is based on the awareness that a qualified concentration of resources and services contributes to effective support of innovation programs and encouragement of research/industrial cooperation, which increases the quality of production and researcher skills (Smilor, Kozmetsky, & Gibson, 1989).

With reference to the technological capabilities, marked by the coming availability in three to five years of ISDN networks and subsequently of Integrated Broadband Communication Networks, the NETT initiative is positioned in the European development framework and linked to the EEC programs for:

- the promotion and development of a European information market; and
- the promotion of advanced telecommunication services (STAR) in the less developed regions of Europe.

NETT is tailored to the particuar need of diffusing advanced telecommunications services in southern Italy as a means for overcoming the region's comparatively slow economic development. In fact, the NETT initiative aims at stimulating the creation and the development of a new business culture in southern Italy, primarily by small- to medium-sized enterprises that presently are not aware of the technical and economic implications of accessing these new communication services.

NETT's realization is not based on an explicit demand, but, on the contrary, is intended as an opportunity and a condition to stimulate latent regional economic needs and to inspire new ones. This is to be accomplished by providing advanced installation and resources and through a whole set of actions (information, promotion, and training) aimed toward potential users. NETT's objective is to contribute specifically toward innovating and strengthening the economic structure of the region and giving it an international position. This plan has been put into action on a local basis by launching specific initiatives, supported by the Italian Ministry for Southern Development. These action initiatives deal with:

(1) the organization and step-by-step upgrading, around the NETT technology, of advanced resources (computing, communications, logistic, and professional);
(2) the installation of high-tech telecommunications infrastructure both within the Tecnopolis Novus Ortus basin and in the other southern Italian regions, to speed the integration process of the local communities into the European network system;
(3) the design, set-up, and marketing of high value-added services; and
(4) the organization of demo training and promotional activities for stimulating the users' demand for new services and for assisting business operators in their growth in an information economy.

NETT and Regional Economic Development

The whole package of initiatives being carried out at Tecnopolis Novus Ortus is aimed at strengthening the role and the capability of the research park as a center for:

(1) the provision of innovative services in the information and communication fields for the benefit of various classes of users (primarily small- and medium-sized enterprises); and
(2) the incubation of new businesses related to telematics services through the training and assistance of local operators.

The technical requirements of NETT stem from both present and foreseen applications/services such as (1) advanced communications (message exchange, document transfer, and conferences among regional users, institutions, and research parks); (2) tele-education and tele-training; (3) tourism applications; (4) image-based tele-medical applications; and (5) geographical information systems. The capability for handling and transferring multimedia information at different speeds and in an interactive way is of outstanding importance. Accordingly, a suitable mix of transmission technologies is being considered, including the use of satellites and fiber optics.

The development of NETT is being implemented with two main objectives: first, the multifunctionality of the Tecnopolis park is characterized by a variety of entities, structures, and resources distributed in the area (such as multimedia applications, multiple interests, and relationships); second, to satisfy the multiplicity and variety of users for the need of connecting to a variety of outer networks and services, and for testing and demonstrating new technological solutions.

NETT Technical Specifications

At present, NETT communications infrastructures provide a high-performance communication system, a satellite ground station, and connections to regional and wide area networks. The communication system is based on a broadband multiservice LocalNet, which extends across the Tecnopolis park (1,900 meters of coaxial cable) and supports both data and TV traffic, including independent TV channels and multiple 128-Kbps data channels (up to an aggregate capacity of 15.4 Mbps) which are frequency-division multiplexed over the cable. The system allows the interoperability of all the TV equipment and the data resources (over 300 terminations) available within the park: mainframes and their computing facilities, workstations, personal computers and terminals, gateways to outside networks, telex/teletex computer-based stations, X25 PADs, and modems on PSTN.

The satellite ground station comprises a 6-meter R/T antenna (operating currently at 11-14 GHz, and soon at 20-30 GHz frequencies) for accessing such satellites as ECS, ECS/SMS, OLYMPUS, and ITALSAT (the Italian domestic satellite). It supports researchers and demo services using video and audio conferences, still-frame video conferences, electronic mailing, high-speed data transmission, tele-education, and tele-training.

Finally, NETT communication infrastructures provide connections to regional and area-wide value-added networks. Tecnopolis Novus

Ortus acts as a nodal center for this network, with links to surrounding areas and to multiple academic/research networks. The expanded networks at both national and international levels include connections to IATINET, the southern Italy network. IATINET constitutes the telematic strategic infrastructure of southern Italy, and has a specific promotional objective—based on Tecnopolis Novus Ortus—for the diffusion of services.

These services run along two main lines of action that are strictly dependent on each other and are related to the STAR program. On one hand, new advanced communication infrastructures (based on satellites and fiber optics) are being installed, experimented with, and enriched with value-added content; on the other hand, comprehensive programs for the set-up, promotion, and diffusion of innovative services are defined and put into effect based on the use of the new infrastructures.

Advanced Technologies

In order to anticipate the availability of extended broadband networks in Italy and in Europe, and to carry out an early experimentation and evaluation of broadband videomatic services, an Optical Island Project is being started at Tecnopolis Novus Ortus. The park area and the surrounding basin will be equipped with a fiber-optics distribution system and about 100 videomatic terminals (at the starting phase). This communication system will be used for the support of both interactive and diffusive wide-band applications and services to be set up and demonstrated for the benefit of such potential user sectors as industry, tourism, health, and education.

Before the full availability of ISDN/broadband networks in southern Italy, and as a means to reach all users, satellite-based communications systems are being provided as a diffused public service. Tecnopolis Novus Ortus will serve as one of the main poles for the provision of the telematic services to local communities. Both a low-speed (up to 64 Kbps) star-fashioned network and a 2-Mbps SMS-type mesh network are being provided. In the former case, selected user sites are being equipped with small antennas (VSAT) placed at both fixed and mobile stations.

The simulation programs being put into operation are based not only on a network of demonstration centers arranged in the southern regions of Italy (provided with specialized labs and environments for training and promotion activities), but also on traveling units for the support of itinerant programs of promotion and training activities on new telematics services for the benefit of less-served user areas.

Conclusion: A Partnership Strategy

The overall package of value-added services supplied by the above structures includes: specialized local information bases of videotex type; mediated access to information coming from available world data banks; electronic mail; computer-based conferencing and tele-videoconferencing; computer-assisted high-quality electronic publishing; on-line automatic linguistic translations; computer-aided design, engineering, and manufacturing; geographical, tourist, and medical imaging; and computer-assisted tele-education and tele-training. The implementation of the project is based on:

- the strategic ability of Tecnopolis Novus Ortus to establish partnerships with industry and service vendors, in order to modernize not only the end users but also the suppliers of advanced information services; and
- the technical, organizational, and professional capabilities of Tecnopolis Novus Ortus, which are being updated in order to support start-up and subsequent development.

The partnership strategy is summarized in Figure 11.2, which describes the distribution of responsibilities and activities between Tecnopolis Novus Ortus and its partners for any given area of information services. During 1987-1988, it was possible to develop more than 100 contacts with different local government offices and small enterprises and to implement more than 20 cases of adoption and utilization of such advanced information services.

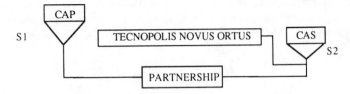

	TECNOPOLIS NOVUS ORTUS	CAS
Responsibility	- Needs Identification -Selection of offerings -Strategy for converting potential demand into market demand - Investment strategy - R&D strategy	-Introduction strategy and tactics -Prototyping -Final defintiion and production of offering -Operations profitability
Action	- Promotion - Informaiton - Generic training - Measure of return investment - Financing - R&D	-Sales -Specific training -Delivery -Installation and support -Updates and improvements -Development for productization

CAP = Commercially available products and packages
CAS = Commercially active service vendors
S1 = Selection of offerings
S2 = Selection of partners

Figure 11.2. Process for Regional Technology Modernization

References

Bozzo, U., Gibson, D. V., Sabatelli, R., & Smilor, R. W. (in press). Technopolis Bari, Italy. In A. Brett, D. Gibson, & R. W. Smilor (Eds.), *University spin-off companies*. Totowa, NJ: Rowman & Littlefield.

Friberg, E. G. (1989, May-June). 1992: Moves Europeans are making. *Harvard Business Review, 3*, 85-89.

Goddard, J. B., & Gillespie, A. E. (1989, November). Advanced telecommunications and regional economic development. *The Geographical Journal, 152*(3).

Huber, R. (1987). Information technology task force. European Economic Community, Brussels, Belgium. *International Journal of Technology Management, 2*(3-4), 501-514.

Magee, J. F. (1989, May-June). 1992: Moves Americans must make. *Harvard Business Review, 3,* 78-84.

Petrella, R. (1989). La mondialisation de la technologie et de l'économie. *Futuribles, 135,* 3-25.

Petrignani, R. (1990, January-February). "Stanford/Italy share knowledge." Keynote address at the first Italian-American Conference on Technology Transfers. *Stanford Observer*, p. 7.

Shurkin, J. (1990, January-February). "$22-million budget cut launched." *Stanford Observer*, pp. 1-20.

Smilor, R., Kozmetsky, G., & Gibson, D. (Eds.) (1988). *Creating the technopolis: Linking technology commercialization and economic development.* Cambridge, MA: Ballinger.

Smilor, R., Kozmetsky, G., & Gibson, D. (1989). Creating the technopolis: High technology development in Austin, Texas. *Journal of Business Venturing, 4,* 49-67.

Snowdon, M. (1988). Moving toward a single market. *International Journal of Technology Management, 3*(6), 643-655.

Tatsuno, S. (1986). *The technopolis strategy.* Reading, MA: Addison-Wesley.

12

Bangalore: India's Emerging Technopolis

ARVIND SINGHAL
EVERETT M. ROGERS
HARMEET SAWHNEY
DAVID V. GIBSON

This chapter describes and analyzes the growth of high-tech microelectronics industries in Bangalore, an emerging technopolis that often is referred to as "India's Silicon Valley." The context of technology transfer in Bangalore's (and, more generally, India's) R&D organizations and private industries is analyzed,

AUTHORS' NOTE: The present chapter is based on research visits to Bangalore to analyze the growth of high technology industries. Interviews were conducted with 70 officials involved in Bangalore's high-tech development, including officials from the central and state government, academic institutions, R&D institutes, private industry, financial institutions (venture capital firms), and electronics entrepreneurs (Singhal & Rogers, 1989b; Singhal, 1989). This chapter also benefits from interviews with managers, scientists, and officials of Texas Instruments, the Indian Consulate in San Francisco, California, the Silicon Valley Indian Professionals Association, and more than 30 other high-tech companies in the Silicon Valley and Orange County area. The research was funded by the Annenberg School for Communication, University of Southern California, and the University Research Institute at the University of Texas at Austin.

and lessons learned about technology transfer in Bangalore are discussed. Arvind Singhal is Assistant Professor in the School of Interpersonal Communication at Ohio State University. Everett M. Rogers is the Walter H. Annenberg Professor of Communications at the University of Southern California, and a Senior Research Fellow at the IC² Institute at the University of Texas at Austin. Harmeet Sawhney is a Ph.D. candidate in the College of Communication at the University of Texas at Austin. David V. Gibson's institutional affiliations are presented in the introductory chapter of this book.

Bangalore: India's Silicon Valley

Several Third World countries—such as Singapore, South Korea, Taiwan, Hong Kong, Brazil, Mexico, Egypt, and India—are attempting to create indigenous high-technology microelectronics industries.[1] Microelectronics, that part of the electronics industry centering on semiconductor chips and their applications (such as in computers and telecommunications), usually is considered the highest of high technology. Some enthusiastic observers claim that microelectronics potentially represents a type of industry that can allow Third World nations to leapfrog the industrial era to become information societies (Singhal & Rogers, 1989a).

The government policies of former Indian Prime Minister Rajiv Gandhi, an airplane pilot and a technophile, promoted indigenous businesses in microelectronics, telecommunications, computers, and computer software. In five years (1984-1989), the number of computers in India increased fifteenfold, computer software exports increased eightfold, and the computer industry's revenues increased fivefold (Singhal & Rogers, 1989b).

Bangalore, the capital of the state of Karnataka in southern India, is the nation's fastest-growing city (8% annually) with a population of four million people in 1989 (Figure 12.1).[2] Large and small microelectronics-based industries increasingly are agglomerating in Bangalore. *Agglomeration* is the degree to which some quality is concentrated spatially in one area (Rogers & Chen, 1988). In 1989, Bangalore housed 375 large and medium-scale industries, of which 135 (36%) were

Figure 12.1. Location of Bangalore in the Interior of India's Southern Peninsula

electronics companies (Vyasulu, 1987). In addition, more than 2,600 small-scale electronics companies operated in the area (Matthai, 1987).[3] In 1989, Bangalore's eight large industrial parks (including the Peenya Industrial Estate, the largest in India) housed an estimated 3,000 companies engaged in electronics manufacturing or assembly, a number steadily growing at the present time. Bangalore's 3,000 large, medium, and small-scale electronics companies employed an estimated 100,000 people, registering sales of $1.2 billion (U.S.) in 1989 (Varma, 1989).[4]

Cumulative Number of Technologies

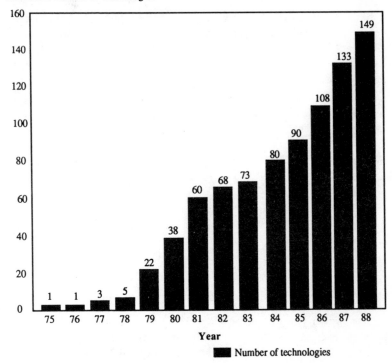

Figure 12.2. Technologies Transferred from ISRO to Indian Industries

SOURCE: Indian Space Research Organization (1989).

Technology Transfer in Bangalore

Bangalore's high-tech R&D organizations conduct both basic and applied research. Although several of Bangalore's R&D organizations actively transfer technological innovation (developed in their laboratories) to private industry, others are relatively slow.

ISRO: A Technology Transfer Success Story

The Indian Space Research Organization (ISRO), headquartered in Bangalore, has a spectacular record in transferring technological innovation to Indian industries. A Technology Transfer Center was established at ISRO in 1982 and by early 1989 ISRO had transferred technical know-how for 160 products and processes to private industry (Figure 12.2), and technology transfer arrangements for another 150 technological innovations were in progress (Indian Space Research Organizations, 1989).

ISRO's technology transfer activities are highly profitable: In 1988-1989, ISRO transferred technical know-how worth $100 million dollars to Indian industries, earning 50% of its annual budget. ISRO has transferred technological innovation to industry in high-precision optics, microelectronics, adhesives, ceramics, computer software, and television hardware (Nilekani, 1988). Several Bangalore-based entrepreneurs started new high-technology companies centered around ISRO's technological innovations. Although many of ISRO's technology transfer arrangements have spurred high-tech activity in Bangalore, created jobs and wealth, and led to import substitution, relatively few have found big commercial markets.

IISc's Center for Scientific and Industrial Consultancy

The Indian Institute of Science (IISc) is Bangalore's (and India's) premier research university, excelling in high-technology areas such as electrical engineering, computer science, material science, aeronautical engineering, and biotechnology. Established in 1909 (by visionaries such as J. N. Tata and Maharaja Wodeyar IV), IISc in 1989 had 30 scientific departments and centers, 400 faculty members, and 1,400 students. During the past 80 years, IISc has trained several thousand scientists and engineers, many presently employed in Bangalore's high-technology industries. Several of IISc's faculty and alumni (like Dr. C. V. Raman, Dr. Homi Bhabha, and Dr. Vikram Sarabhai) helped found many of Bangalore's scientific institutions and R&D labs.

Through the Indian Institute of Science's Center for Scientific and Industrial Consultancy (CSIC)—established in 1973—industry-sponsored R&D is undertaken, and technological innovations developed in IISc's labs are transferred to Bangalore's private industries. By 1989, more than 2,000 projects (ranging from one-day consultancy projects to complete technology transfer arrangements) were completed by CSIC. By 1989, 180 (40%) of IISc's 450 faculty members had participated in CSIC-industry projects. IISc faculty are allowed to consult with industry the equivalent of one workday per week and retain 50%

of the consultation fee; the remaining 50% goes to IISc. IISc's policies on faculty consulting are relatively liberal compared to other scientific institutions in India, an important reason for CSIC's success.

CSIC has been proposed as a model center for academic-industry collaboration, especially for the five Indian Institutes of Technologies (IITs), which are excellent teaching universities but are low in research output and have weak links to industry (Singhal & Rogers, 1989b). Although IISc is understandably proud of its excellence in research, several factors limit the impact of IISc's technology transfer activities. Most research conducted at IISc is basic, as opposed to applied research. Also, IISc's CSIC has maintained a somewhat low profile in Bangalore. In 1988-1989, CSIC received 350 project proposals, of which only 150 were undertaken (Subramanya, 1988). Furthermore, CSIC is understaffed severely for its wide-ranging activities. Ways to boost applied research at IISc while maintaining ongoing basic research and further strengthen the IISc-industry interface are being explored.

C-DOT: A Model R&D Organization

In 1984, Satyan ("Sam") Pitroda, an overseas-returned Indian (formerly an executive of Rockwell, Inc., in Chicago, and holder of 50 patents in telecommunications equipment), founded the Center for Development of Telematics (C-DOT) headquartered in Bangalore and New Delhi. C-DOT represents an organizational model for Indian R&D institutes: It conducts state-of-the-art R&D to meet Indian's specific telecommunication needs; is functionally autonomous from the government; has a dynamic leader in Sam Pitroda, a flat hierarchical structure, and highly motivated goal-oriented scientists and engineers; and it transfers technological innovations (developed in its labs) to private industry (Singhal & Rogers, 1989b).

C-DOT developed state-of-the-art telephone switching equipment to serve India's special telecommunications needs, which are high traffic and low density (as compared to low traffic and high density in most Western countries) and extreme temperature and humidity conditions. C-DOT successfully developed the technology for (1) electronic PABX systems, (2) a 128-line rural automatic exchange (RAX), and (3) 4,000-line and 10,000-line main automatic exchanges (MAXs). In 1986, C-DOT's EPABX technology was transferred to 42 private vendors, several of which established manufacturing operations in Bangalore (for example, Unitel Limited). In 1988, C-DOT's RAX technology was transferred to Indian Telephone Industries (ITI), and production of RAXs began in a C-DOT–ITI plant located in Bangalore's Electronics City.

Ancillaries and Technology Transfer

Bangalore is home to six large public sector companies, of which several conduct high-tech defense-related work: (1) Indian Telephone Industries (manufacturer of telecommunications equipment, electronic exchanges, and integrated circuits); (2) Bharat Electronics Limited (a government defense contractor); (3) Hindustan Aeronautics Limited (HAL) which began during World War II as a repair/maintenance center for Allied fighter planes, and presently produces supersonic MIG aircraft; (4) Hindustan Machine Tools (HMT), producer of state-of-the-art machine tools and precision watches; (5) Bharat Heavy Electricals Limited (HBEL); and (6) Bharat Earth Movers Limited (BEML), which spun off from HAL in 1964. Encouraged by the government's industrial policy of self-reliance, some 130 ancillary companies have been established by ITI (48), BEL (22), HAL (10), and HMT (50), most of which are agglomerated in Bangalore (Rao, 1987).

The ancilliary industries represent a potential conduit to transfer technology from the large public sector companies to the commercial marketplace. The ancilliary industries in Bangalore, however, are generally captive suppliers to one or few of these large public-sector companies, thus reducing the market potential of their products. Although the large public-sector company buys the ancilliaries' products, the transferred technological know-how (of producing these products) does not get commercialized and introduced into the marketplace.

Understandably, Bangalore's small companies (ancilliaries included) fare worse when it comes to international technology transfer. Quite often the resources that are required to span vast distances are limited. Only large high-tech companies have the staying power and resources to risk overseas ventures. Even in the case of computer software, small companies lack start-up capital, market research, and the staying power to compete with bigger companies.

Critical Mass, Venture Capital, and Technology Transfer

Agglomeration of high-tech firms in a technopolis is important because it concentrates a critical mass of successful entrepreneurs, who then support and reinforce each other in the uncertain situation of launching new companies. The concept of a critical mass comes from nuclear physics, where it refers to the minimum amount of radiation needed to set off a chain reaction (Rogers & Larsen, 1984). In nuclear physics a chain reaction is set off only when fissionable materials of a certain grade of purity are brought together. Impurities, if present, dampen the interaction between fissionable particles and thereby prevent a chain reaction.

In Bangalore, the level of interaction between co-present elements of the critical mass is not very active in transferring technology between R&D establishments and private industry. Because such interaction is essential for Bangalore's further development as a technopolis, ways to enhance a self-sustaining chain reaction of technology transfer must be found. Bangalore's high-tech environment can be enhanced by such efforts as infrastructure development (for example, an improved local telephone service), government support (for example, tax incentives for high-tech firms), and financing of entrepreneurship (for example, the presence of venture capital). For balanced, coordinated high-tech growth, cross-institutional cooperation is necessary (Smilor, Gibson, & Kozmetsky, 1989).

The role of venture capital is crucially important in fueling high-tech growth (leading to a "chain reaction") in a technopolis (Rogers & Larsen, 1984; Segal, Quince, & Wicksteed, 1985; Singhal & Rogers, 1989b). Most entrepreneurs are technologists, not businessmen, and so the venture-capital firm makes an important contribution to the entrepreneurial start-up by providing managerial advice to the entrepreneur. Most Indian technologists (also true in Bangalore) come from middle-class families which, although highly educated, normally do not have access to large amounts of private capital. The main source available to an entrepreneurially driven engineer are banks, which are usually resistant to funding high-tech start-ups.

Until 1988, venture capital for high-tech start-ups was virtually nonexistent in Bangalore. In 1988, India's first private venture-capital company, Technology Development and Information Company of India (TDICI), established its headquarters in Bangalore. TDICI's president, P. Sudarshan, served earlier as chairman of ISRO's Technology Transfer Center in Bangalore. In 1989, TDICI invested $13 million (U.S.) in 40 high-technology ventures, with several located in Bangalore (Staff, 1989). In 1990, TDICI appraised another 90 high-technology ventures, representing a capital investment of $31 million (U.S.).

Although venture capital is a recent phenomenon in Bangalore, it is gaining momentum thanks to TDICI, Karnataka State Industrial and Investment Development Corporation (KSIIDC), and Karnataka State Financial Corporation (KSFC), two state-level financial institutions. KSIIDC and KSFC invest in large, medium, and small-scale industries (including several high-tech ventures), often as equity partners, and provide technomanagerial support for start-ups. KSIIDC (in cooperation with the Department of Electronics, New Delhi) has established a computer software park in Bangalore.

Reverse Technology Transfer: The Case of UNIX

In the past decades, most international technology transfer has been one-way—from Western industrialized countries to the Third World. The transfer of UNIX know-how from India to Western countries (primarily the United States) represents a spectacular case of technology transfer in the reverse direction.

After the exit of IBM from India in 1978 for refusing to allow more than 50% Indian ownership, the government of India evaluated various standards for the indigenous computer industry and decided on UNIX because of its nonproprietary nature. Although Western computer companies used UNIX mainly for educational and scientific purposes, Indian engineers developed UNIX for more wide-ranging commercial applications. Subsequently, as UNIX became more popular worldwide, Indian engineers with expertise on UNIX became highly sought. Several Indian firms with UNIX expertise are based in Bangalore, and many have secured UNIX export contracts from U.S. companies. Hindustan Computers Limited (HCL), India's largest computer manufacturer and a leader in UNIX, established a Silicon Valley-based subsidiary (HCL, America) that conducts UNIX-based R&D in India and subsequent manufacturing in the United States. HCL, America is one Indian company—among many others—that is a leader in reverse technology transfer.

Texas Instruments: A Case Study of
Two-Way International Technology Transfer

The rapid growth of the computer industry has led to an acute shortage of software engineers worldwide. India, with 3 million scientists and engineers, is uniquely placed to capitalize in the worldwide computer software markets. Furthermore, Indian engineers fluent in English are available at one-tenth of the cost of comparable talent in Western industrialized countries. Although India's initial response to global software opportunities was to export engineers (to work at clients' facilities overseas, primarily in the United States), currently India is exporting software products and services in addition to skilled workers and scientists.

One apparently successful strategy for boosting India's domestic computer software industry has been to use telecommunications links to access foreign markets. In pioneering such an effort, Texas Instruments (TI) established a software development center at Bangalore that is linked via a dedicated satellite to its Dallas headquarters.

When G. R. Mohan Rao—a vice president of TI and an expatriate Indian—accompanied TI's president to India in 1983, both were impressed favorably by the high caliber and low cost of Indian engineers, the attractive climate of Bangalore, and the political situation in New Delhi that favored high-tech development. In an innovative and bold move, TI decided to establish a software design facility in Bangalore. Government approval was secured in 1985. In 1990, TI employed 120 Indian engineers (mainly from IISc, IITs, and Bangalore's engineering colleges). TI engineers in India are paid a starting salary of about $250 (U.S.) a month, reasonable by domestic standards, but only about one-tenth of what TI would pay equivalent employees in the United States.

TI's software facility in Bangalore represents an investment of $5 million (U.S.) and is 100% U.S.-owned. In return, the Indian government requires TI's Indian business operations to be 100% export-oriented: In return for every dollar of foreign exchange earned, the Indian government allows TI to import into India (without any customs duty) sophisticated computer hardware and software packages of value equal to the exports. TI Bangalore researchers carry out computer-aided design of VLSI circuits; the research process is linked to TI Dallas by satellite, which transmits the software code.

India is trading brainpower for TI's technology in what appears to be a unique case of two-way international technology transfer, representing a win-win situation for both parties. TI has gained a foothold in India; Indian engineers have a foothold in TI; India is retaining its talent and getting it trained at TI's expense; the Indian government earns foreign exchange; and TI's Indian engineers are happy, learning state-of-the-art VLSI design in a pleasant work environment.

Software development in India goes beyond maintenance of "software factories" or "sweatshops" where labor-intensive work such as data entry, porting operations, digitalization of drawings, and typesetting are conducted. Although such business is welcome in India, more lucrative opportunities lie in software development work. Only about 30% of TI's Bangalore operations comprises routine work; 70% is developmental. Such developmental effort is not directed toward new data base concepts or new operating system designs but rather toward developing research ideas into prototypes, a high-risk venture with no guarantee of success. Because of India's lower operating costs, the financial risk in development projects is reduced drastically. India is thus an excellent site for development of "wish lists" or "zero budget" projects. For example, Indian programmers took 35 man-years to develop Instaplan, a highly successful computer software product, at about one-tenth the cost of what the project would have cost in the United States.

India offers several other advantages to TI: Indian maintenance of hardware has been better than expected; Indian workers have adapted easily to TI's work culture; the Indian government has been responsive to TI's needs; and time differences between TI India and TI Dallas permit continuous work around the clock.

TI has also found, however, that satellite communication does not substitute entirely for face-to-face interaction, especially when developing complex computer software codes. The success of such long-distance R&D depends much on the initiative and caliber of TI's Bangalore engineers. TI has overcome the distance barrier partially with the combination of an egalitarian work culture (which encourages individual initiative) coupled with the high-quality training and self-motivation of Indian engineers. The success of TI's pioneering effort has inspired other similar ventures. For example, plans are underway to establish a satellite link between Route 128 companies in Massachusetts and a software development park near Pune, India. The Japanese are planning a satellite-linked software development park near Nagpur in central India. And other U.S. high-tech firms, such as Hewlett-Packard and Digital Equipment Corporation (DEC), have followed TI to Bangalore.

Technology Transfer by Employee Attrition

When Texas Instruments sought approval to establish operations in India, the government of India stipulated that TI not recruit engineers already working in the Indian computer industry. The government feared that TI's higher salaries would siphon off the brightest engineers from Indian computer companies. To allay such fears, TI hired only recent engineering graduates from Indian universities and research institutes. These initially hired employees were sent to Dallas, Texas, and Bedford, UK (where TI has another software facility), for advanced training. They have since become the core for TI's India operations, and subsequent recruits have been trained in India.

Several Indian engineers (about 10%) have left TI to work for other Indian computer companies, and a few have even spun off their own computer software companies. The company expects more engineers to leave in the future. A director of TI's Indian operations stated: "With the passage of time, the attrition of our technical staff will make a general contribution to the transfer of technical know-how to Indian industry." Our discussions at TI, Bangalore, showed that family ties provided an important impetus for engineers to leave TI. Engineers from all over India join TI for outstanding professional opportunities, a pleasant work environment, and good remuneration. But soon, family "pull" leads them to find jobs in their home cities. Thus, through employee attrition TI technology and training is transferred into the Indian high-tech industry.

Technology Transfer by Technological Presence

Another form of technology transfer from TI's Bangalore operation occurs when TI's sophisticated computer workstations are serviced locally in India. For example, TI's Apollo workstations are serviced by Hindustan Computers Limited (HCL), India's largest computer company. By servicing TI's equipment, HCL acquired maintenance expertise and familiarity with the state-of-the-art Apollo workstation. In fact, beginning in 1989, HCL started manufacturing Apollo workstations in India.

Bangalore: A Regional Economy with Global Ties

The case of Texas Instruments in Bangalore illustrates that high-tech areas in developing countries more often are integrated into global markets than the national economy. Many of Bangalore's emerging high-tech companies are integrated more into overseas markets than into the national or even the regional economy (Gollub et al., 1988). This globalization is particularly true for Bangalore's software companies. In this sense, Bangalore is "a foreign land within India."

Should the policies of the Karnataka state government (headquartered in Bangalore) facilitate further concentration of resources within Bangalore, making it a high-tech island with links primarily with global markets? In India, such a scenario not only is difficult to justify politically, but also raises important questions of regional socioeconomic inequality (Rogers & Sawhney, 1989). Another strategy would be to allow further concentration of resources in Bangalore, but at the same time encourage better economic and technological links with the hinterland. More developed domestic linkages would further diffuse the economic gains arising from Bangalore.

The infrastructure to enhance domestic and international communication is vital to reap benefits from regional economic development (such as that occurring in Bangalore). The Indian government has created NICNET (National Information Center Network), a versatile computer network, which makes electronic information transfer within and among 439 administrative districts. NICNET also will have nodes in Boston, Washington, D.C., and in Silicon Valley. The government is aiding linkages between Indian entrepreneurs by connecting major cities through Telecommunication Research Centers' packet-switching network, and is providing access to international telecommunications networks through the Indian National Satellite (INSAT).

Unsuccessful Technology Transfer: The Case of LCDs

For every successful technology transfer effort between Bangalore's R&D institutes and private industry, there exist many failures. One such failure (which proved to be a costly one) was the Indian-based attempt to transfer liquid-crystal display (LCD) technology from the Raman Research Institute (RRI) in Bangalore to Bharat Electronics (BEL), also headquartered in Bangalore.

In 1975, Dr. Chandrashekar perfected the LCD technology in his RRI lab in Bangalore. Although Japanese and U.S. scientists were close on Chandrashekar's heels in developing this technology, commercial worldwide exploitation of the LCD (in pocket calculators, watches, etc.) was yet to begin. RRI transferred its technical know-how to BEL, which in turn produced several prototypes of LCD watches, displaying the tremendous commercial potential of RRI's technology. For five years (between 1976 and 1981) BEL pleaded with India's central government in New Delhi to approve a large LCD manufacturing plant, but the New Delhi bureaucrats provided no leadership. Meanwhile, Japan and the United States were quick to commercialize LCDs, reaping several billion dollars worth of sales a year. In 1984, the Indian government approved a technology transfer arrangement between Hitachi of Japan and Semiconductors Complex Limited (India) to produce LCDs, completely overriding the indigenous LCD know-how that had existed at RRI.

Integrated Technology Transfer: U.S. AID Center for Technology Development

Since 1987, the New Delhi office of the U.S. Agency for International Development (AID) has initiated programs to strengthen the market-driven R&D needs of Bangalore's industries, focusing especially on such high-technology industries as microelectronics, telecommunications, computers, software, and biotechnology (Gollub et al., 1988). U.S. AID's interest to boost high-technology industries in Bangalore was heightened during 1987-1988, when AID successfully implemented two technology-development programs in India: PACT (Program for the Advancement of Commercial Technology) and PACER (Program for the Acceleration of Commercial Energy).

With input from the Karnataka state government, the Indian Institute of Science, Bangalore's R&D institutes, TDICI (the venture-capital company), and private industries, U.S. AID is establishing a Center for Technology Development (CTD) in Bangalore to strengthen technology transfer between R&D centers and industry, and the subsequent commercialization of technological innovations (Arthur D. Little, Inc., 1987).

The idea of CTD grew out of a recommendation by *Karnataka in Transformation,* an AID-sponsored study, which observed that:

A buyer-supplier initiative extends the chain of technology down from the second-generation level to the first-generation level. Applied technology may make Karnataka firms more innovative and able to compete, but moving the technology they are using to the supplier is equally essential. (Gollub et al., 1988, p. 167)

As an applied research and development center, CTD will coordinate activities of Bangalore's major high-tech players (R&D institutes, state government, private industry, and venture-capital firms), helping with (1) technology selection, (2) technology development, (3) technology transfer, and (4) technology commercialization initiatives. CTD's main focus will be to enhance buyer-seller relationships.

Lessons Learned

What lessons can be learned about technology transfer from such cases in Bangalore as ISRO, IISc, C-DOT, ancilliary industries, TDICI, UNIX, Texas Instruments, RRI-BEL, and U.S. AID's Center for Technology Development?

(1) R&D institutes can make heavy profits (through license fees, etc.) by transferring technological innovations to private industry. ISRO in Bangalore earns about half of its annual R&D budget by transferring technologies to private industries.

(2) R&D institutes that conduct high-tech defense research (and have classified security status) can facilitate the transfer of their nonmilitary technological innovations to private industry by establishing autonomous technology transfer entities; for example, ISRO's establishment of an autonomous technology transfer corporation.

(3) Technology transfer can be facilitated through sharing of R&D facilities, people exchange, and faculty consultation, as illustrated by the activities of IISc's Center for Scientific and Industrial Consultancy.

(4) R&D institutes need to find a balance between basic and applied research. For example, C-DOT conducts basic and applied research in the area of telecommunication technology, representing an organizational model for Indian R&D institutes.

(5) Some of the best technology transfer occurs when individuals move from one location to another. For example, Sam Pitroda's return to India brought home not just the individual, but also state-of-the-art telecommunications technology that Pitroda helped develop in the United States.

(6) Ancilliary industries represent a potential vehicle to transfer technology from large high-tech companies to the commercial marketplace. Not much of this potential has been realized to date, as Bangalore's ancilliaries are generally captive suppliers to large high-tech companies.

(7) Ways to enhance the purity of Bangalore's high-tech environment must be found to trigger a self-sustaining chain reaction in behavior's critical mass of companies. The availability of venture capital (provided by TDICI and state financial institutions), and of technomanagerial support to entrepreneurs are crucially important in fueling high-tech growth in a technopolis.

(8) Technology transfer need not always be from Western industrialized countries to Third World countries. The Indian expertise in UNIX represents a spectacular case of reverse technology transfer, from a Third World country to Western industrialized countries.

(9) Technology transfer can go beyond the mere transfer of a hardware technology or individuals, and include transfer of organizational structures, work environments, and managerial innovations. For example, Indian engineers at TI's Bangalore facility work in an American-style management system, much like their counterparts in Dallas.

(10) Advances in communication technology—for example, the availability of a dedicated satellite link for data transfer (as in the case of TI)—minimize the barrier of geographical distance in the international transfer of brainpower. Communicating via a satellite link, however, does not substitute entirely for face-to-face interaction, especially when developing complex computer software codes.

(11) Employee turnover and attrition is one effective way of technology transfer in high-tech firms. Engineers who leave high-tech companies to join others or to spin off their own companies carry with them the technical know-how learned in their parent companies.

(12) An apathetic, inefficient bureaucracy can be anathema for technology transfer efforts, especially when the lead time to exploit a technological innovation commercially is small. For example, the RRI-BEL project on LCDs got nowhere because of apathy from India's central government in New Delhi.

(13) Indigenous technology development and technology transfer arrangements compete with options of imported foreign technology. National governments and private industries need to weigh carefully the pros and cons of developing indigenous technologies versus the option of importing ready-made foreign technologies.

(14) Synergy between participating organizations (R&D institutes, government agencies, private industries, venture-capital firms, and entrepreneurs) is important in the transfer of technology. An umbrella organization, like the U.S. AID-sponsored Center for Technology Development in Bangalore, can serve as a coordinator and catalyst in the technology transfer process.

Notes

1. A high-tech industry is also characterized by (1) a high proportion of highly-skilled employees, many of whom are scientists and engineers, (2) a fast rate of growth, (3) a high ratio of R&D expenditures to sales (typically about 1:10), and (3) a worldwide, highly competitive market for its products (Rogers & Larsen, 1984).

2. High-technology growth also can be found in other urban centers such as Pune, Chandigarh, Delhi, and Bombay.

3. Indian industries are classified as large, medium, and small based on the amount of start-up capital investment: large-scale industries represent an initial investment of more than $2.2 million (U.S.); medium-scale industries represent an initial investment of between $220,001 to $2.2 million; and small-scale industries represent an initial investment between $16,250 to $220,000.

4. Sales figures are computed at an exchange rate of 16 rupees to one U.S. dollar.

References

Arthur D. Little, Inc. (1987). *Technology development on a state level focused on national goals*. New Delhi: Arthur D. Little, Inc.

Gollub, J., Hansen, E., Gorbis, M., Krishna, S., Puri, A., & Waldhorn, S. (1988). *Karnataka in transformation: A blueprint for action*. Menlo Park, CA: SRI International.

Indian Space Research Organization. (1989, March). *Indian space programs partnership with industry*. Bangalore: Indian Space Research Organization.

Matthai, P. (1987, October). *Bangalore's medium and small scale industries: Future perspectives for development in the intra-state regional context*. Paper presented to seminar on Bangalore 2000—Some Imperatives for Actions Now, Times Research Foundation, Bangalore.

Nilekani, R. (1988, July 24-30). Space industry: A high-tech spinoff. *Sunday,* pp. 74-75.

Rao, K. K. (1987). *Public sector in Bangalore's metropolitan economy: The 2000 A.D. perspective*. Paper presented to seminar on Bangalore 2000—Some Imperatives for Actions Now, Times Research Foundation, Bangalore.

Rogers, E. M., & Sawhney, H. (1989). *Institution-building aspects of high-technology*. New York: United Nations Development Program, Central Evaluation Office.

Rogers, E. M., & Chen, Y. A. (1988). Technology transfer and the technopolis. In M. A. V. Glinow & S. A. Mohrman (Eds.), *Managing complexity in high-technology industries: Systems and people*. New York: Oxford University Press.

Rogers, E. M., & Larsen, J. K. (1984). *Silicon Valley fever: Growth of high-technology culture*. New York: Basic Books.

Segal, Quince, & Wicksteed. (1985). *The Cambridge phenomenon*. Cambridge, England: Segal, Quince, Wicksteed.

Singhal, A., & Rogers, E. M. (1989a). A high-tech route to development? *Interaction, 7*(1).

Singhal, A., & Rogers, E. M. (1989b). *India's information revolution*. Newbury Park, CA: Sage.

Singhal, A. (1989, December). *Bangalore: Lessons learned from India's emerging technopolis*. Paper presented to Pyramids Technology Valley Symposium, Cairo, Egypt.

Smilor, R., Gibson, D., & Kozmetsky, G. (1989). Creating the technopolis: High technology development in Austin, Texas. *Journal of Business Venturing, 4*, 49-67.

Staff. (1989, July). Projects assisted by TDICI. *Venture India.*

Subramanya, M. (1988, July 10). Industrial application of research. *The Economic Times,* pp. 3, 6.

Varma, K. (1989, May). Karnataka electronics: Rung by rung to the top. *Telematics India,* 66-91.

Vyasulu, V. (1987, October). *Industrial scenarios for Bangalore.* Paper presented to seminar on Bangalore 2000—Some Imperatives for Actions Now, Times Research Foundation, Bangalore.

13

Multinationals: Preparation for International Technology Transfer

EUN YOUNG KIM

As multinational firms have been recognized as principal sponsors of international technology transfer, researchers and practitioners increasingly have been interested in the role of people in these transfers. People exchange technical information when they meet through exchange programs, transfer programs, technical conferences, and consultation services. The centralized training programs of multinational corporations are a good example; personnel from different countries gather at one site to learn new technologies and take them back to their home countries. In this study, Eun Young Kim examines the role of multinational consulting firms and their corporate training in international technology transfer. The situation is elaborated in a brief case study of Andersen Consulting, by whom Dr. Kim was formerly employed. Currently, he is President of Vision 2000—a communication consulting firm in Seoul, Korea.

The Move Overseas

The rapid development of information technology has had a profound impact on today's worldwide businesses. The information revolution, along with the globalization of the economy, has contributed to the expansion of international markets. In this milieu, a great number of U.S. firms have moved overseas. Seeking worldwide leadership in their respective industries, U.S. firms have increased their use of management and consulting services. Consequently, U.S. consulting firms also have been moving beyond their national boundaries to meet their clients' demands (Nees, 1986). As of 1985, there were 600-700 U.S. management and information consulting firms operating in over 100 countries (Stiffler, 1985) and the number has been increasing continually.

The growth of the international consulting market is shown in the revenues and the numbers of employees at major management and information consulting firms of the United States. For example, in 1988 the 10 largest U.S. management and information consulting firms had worldwide revenues ranging from $1,120 million (Andersen Consulting) to $375 million (Saatchi & Saatchi). In terms of the increase of employees in the consulting profession, Strizich (1988) reports a 450% increase within the last decade. In fact, Andersen Consulting tripled their consultants from about 6,000 in 1984 to approximately 18,000 in 1989. Indeed, major consulting firms have been pressuring themselves toward dominance in the international consulting business, intensifying their efforts to expand target clients to include local business, industry in other countries, and U.S. subsidiaries.

Whether the main objective of international consulting is to transfer technology or not, a major element of international business is the transfer of technology from one country to another (Frame, 1983). For example, when a Korean firm purchases a license to use an American firm's system-installation methodology, technology transfer is the explicit objective of the transaction. On the other hand, in some cases technology transfer is merely a side effect of the purchase of goods (Frame, 1983). For instance, a Thai consumer buying an IBM computer is also purchasing the technology embodied in the device, although technology was not the first objective. In either case, however, technology is being transferred across national boundaries.

International Technology Transfer

Process of Technology Transfer

Although the definitions of technology transfer vary, there seems to be agreement about the major elements of the technology transfer process. The fundamentals include a donor, a recipient, and a transfer mechanism. In much international technology transfer, the donors are U.S. multinational firms and the recipients are less developed countries (LDCs), including developing countries and newly industrialized countries (NICs). The transfer mechanisms examined in the present chapter focus on the transfer of capabilities rather than the sale of products, including such mechanisms as formal training, on-the-job training, and personnel exchange.

Donor: Multinational Firms

In the past, technology transfer by multinational firms often took place through: (1) the sale of licenses to manufacture; (2) associated technical assistance; and (3) co-production in which an enterprise in one country builds a product for another (Frame, 1983). Because detailed records of such business transactions are typically withheld from public view, it is hard to tell how much technology has been transferred among countries. It is reasonable to assume, however, that technology transfer from the United States to less developed countries will continue as more U.S. firms operate globally.

The motives for private industry to transfer technological knowledge to less developed countries include: (1) the stimulation of new growth markets, (2) access to strategic materials, (3) investment protection, (4) moral imperatives, and (5) world power shifts. Although some may disagree with noble motives such as moral imperatives, U.S. private industry has been instrumental in helping LDCs become self-sufficient in developing the industrial technology and infrastructure needed to support the industrial growth.

Recipient: Less Developed Countries

Many developing and newly industrializing countries, in their effort to accelerate industrial growth, still rely on multinational firms as the main source of advanced technology. Consequently, multinational management and information-consulting companies play a significant role in technology transfer. For example, the total market for consulting services in Malaysia was projected to increase by 10% to 12% per year from 1989 to 1991 in line with the expected improvement in the economy. Americans are one of three major factions in the Malaysian management and technical consulting sector, along with the British and the Malaysians ("Malaysia," 1989).

Although technical consulting for less developed countries has been practiced for a long time, over the years these countries have started to recognize the following paradox: technological self-reliance requires their continued dependence on multinationals. What has happened is that in the past transfers occurred principally through the sale of products that embodied proprietary technology, rather than through empowering the work force with capabilities to develop technology. But, as the impact of human capital on national economies has been publicized around the world (e.g., OECD or UNESCO publications), the LDCs have become much more aware of the importance of educating their own people. Consequently, the recipient countries who used to be simple buyers of technological products are seeking ways to improve the capabilities of their own people to develop new technologies as well as to use them. LDCs have learned the importance of know-how implied in a Chinese proverb: "If you give a needy person a fish, he'll come back to ask for more fish tomorrow, but if you teach him how to fish, he'll not have to come back."

Of course, the readiness to receive sophisticated technologies differs from one recipient to another. For example, countries like Korea and Taiwan have grown in their capacities to develop technology and to adopt more sophisticated technology from the developed world. In fact, they are considered to have highly educated work forces, with high percentages of their populations with advanced degrees in engineering and technical fields. Other developing countries, such as Sri Lanka and Nepal, will need more time to prepare their infrastructures for high-tech knowledge.

Strategic Alliances Between Donor and Recipient

As the world economy is getting more global and technologies more complex, cooperation between donors and recipients has become increasingly critical to successful technology transfers. Although the interests of donors may not always correspond to those of recipients, alliances are essential for expanding markets and serving customers in the global market. As Ohmahae (1989) pointed out, operating globally means operating with partners—local government, local business and industry, and local citizens—and that, in turn, means a further spread of technology. Thus, cooperation from local partners becomes critical to multinational firms and their efficient operation in the international marketplace.

Multinationals seek to build positive public images in their local markets. Multinational firms would like to be perceived as beneficial to the local economies in ways such as providing employment opportunities to local people or expanding their technical capabilities. In their effort to regain their seemingly lost economic competitiveness in the global market, U.S. firms have learned that to be successful multinationals must learn to operate in local contexts.

These combined forces have led many multinationals to adopt policies of hiring local citizens for their branches and subsidiaries. In fact, the advantages of hiring local people are multiple: (1) the salaries of local employees are much lower than those of their U.S counterparts; (2) the firms do not have to pay for moving and living expenses of U.S. employees; (3) local governments favor employment opportunities for their citizens; (4) local citizens know the language and culture in which local businesses operate; and (5) the firms do not have to deal with immigration processes. It seems that human resources may be less mobile than technology, especially across cultures (Carnevale, 1989).

Hiring local citizens and training them how to use the respective firms' methods and products have become increasingly important tasks for multinational firms and the need for continuous training—either formal or informal—will become more critical as the pace of technology change accelerates.

Education and Training as a
Transfer Mechanism

Training employers is an investment for quality improvement, product innovation, and client service. From an employees point of view, tailored training for a high-tech jobs (in contrast to more general academic preparation) is appealing because it directly supports their career development and their value to the firm. Some of the most advanced technologies often are being developed and/or used by business and industry rather than academia. Training opportunities are big incentives to lure potential employees to certain employers.

Formal and Informal Training

The trend of corporate training processes has been from informal toward formal learning (Carnevale, 1989). As the pace of economic and technological change has accelerated in the last half century, employers have tried to ensure the efficiency and quality of learning by formalizing learning processes. Employers have managed to maintain the link between learning and real jobs by applying methods that translate real-world learning needs into structured learning processes. In this process, many companies have constructed major training centers. Eurich (1985) found that American corporations were spending $40 billion for employee training.

Formal training programs are either centralized or decentralized, depending on the topics and purposes of the training. For example, a new product, strategy, or technology may require training a large group of employees as quickly and consistently as possible, so employers tend to provide centrally controlled training in the first stage of innovation. Once the innovations are in place, training becomes more decentralized to fit the specific purposes of divisions, individual job categories, product lines, and strategies.

Different companies use different methods, depending on their needs, budgets, geographic areas, and so forth. IBM organizes its education functions for 132 countries into five geographic units (Galagan, 1989). In each unit, there is an education organization that acts as the coordinating unit; then, within each country, there is an education operation. 3M is more decentralized in its approach to international training than IBM. The company (based in St. Paul, Minnesota) has subsidiaries in 52 countries, and each of them is responsible for its own training. Training may also occur on the job without a classroom or instructor. For example, when employees are working in a team, they can learn technical skills from each other on the job.

Instructional Technology

Corporate training processes, especially decentralized training, have benefitted from the development of advanced instructional technology. Although studies have shown that traditional instructor-led, paper-based training yields substantial monetary savings, companies have adopted various advanced instructional technologies to diversify their training formats. They include video-based training, computer-based training, and interactive video training. Each format has advantages and disadvantages. Video-based training has the advantage of low cost and high comfort level, although it is difficult to adapt to individual user needs. Computer-based training allows students to pace themselves, but it can be boring and frustrating. Interactive video instruction blends the strengths of computer and video methods, but its initial production cost can be very high (Goldstein, 1988).

The cost of developing technology-based courses is often higher than that of paper-based courses. Jack Boshwer, former director of external education at IBM, argues that designing some courses for delivery with advanced technology systems should pay off quickly because of the large number of students who could be trained with each course (Galagan, 1989). In fact, since the 1970s, IBM education has been moving steadily out of classrooms into the circuits and airwaves of its increasingly sophisticated delivery systems. About half of its courses currently are classroom-based, and half are technology-delivered. Most students, including IBM customers, can now get to their satellite class-rooms without traveling overnight. Galagan (1989) reports that by 1992 IBM expects to deliver 250,000 student days per year on satellite systems, thereby avoiding to pay several million dollars of expenses each year on instructors, student travel, and housing. By the end of the 1990s, it is predicted that only 25% of IBM education will take place in classrooms.

State-of-the-art technology-based education is not just a corporate phenomenon; traditional education institutions have also started pro-viding media-based education. One innovative example is the National Technological University (NTU) in Fort Collins, Colorado. The NTU, a consortium of 24 engineering schools, operates an instructional tele-vision network (ITV) via satellite for convenient, flexible on-site engi-neering and technical education nationwide. The NTU was founded on the idea that telecommunications can help overcome barriers of geographic distance and limited instructional resources in continuing engineering education. (The lack of instructors to provide appropriate continuing engineering education, however, has been a major problem.)

Using satellite delivery systems, during 1987-1988 the NTU offered 150 advanced engineering courses to engineering and technical profes-sionals in business, industry, and government organizations across the country. The receiving organizations install satellite receiving stations and receive courses originating from regularly scheduled, on-campus classes that are held in specially equipped studio classrooms. Thus, employees at receiving sites can listen not only to lectures, but also to classroom discussion. A variety of signal delivery systems are also employed to facilitate communication between the graduate students at job sites and the campus. Electronic mail, using one of the packet-switched networks, is the primary means of interaction between students and instructors. Computer conferencing is another mechanism of communication through which all students can participate and interact with students at other sites as well as instructors. In addition to making the best use of satellite-transmission time and the working student's class time, many course transmissions are recorded at the students' sites on videotape machines for use at the students' convenience.

Some educational institutions see satellite education delivery systems from an entrepreneurial perspective. For example, Stanford University's College of Engineering is looking into the possibility of delivering technical programs to local engineers and technicians in Mexico, leveraging on the college's success in offering graduate engineering programs to San Francisco-area engineering and technical professionals through its instructional television station. Georgia Institute of Technology is also endeavoring to open its education market in China and Costa Rica. A motivation of both institutions is to generate more income; nevertheless, it is also a powerful mechanism to transfer technology from the United States to less developed countries.

Education, Site Visits, and On-the-Job Training Abroad

Although U.S. companies seldom send their employees to educational institutions overseas (perhaps because the U.S. higher education system is considered one of the best in the world), it is common for foreign companies to send their employees to U.S. graduate schools on both a short-term and a long-term basis. The employees often return to their jobs with improved English, new professional contacts, and innovative ideas (mostly from basic research). Although such knowledge transfers do not always turn into innovative products, a CEO of a Japanese high-tech company revealed that his firm enjoyed a leading position in voice-recognition technology after the company had sent its researcher to study phonetics in the United States. In this case, the major aim of corporate education was met without the construction of new academic building or classrooms.

On-the-job training abroad is another effective mechanism of technology transfer from one country to another. Visiting overseas employees simply observing how others operate a piece of machinery, manage a complex process, or use resources can be a very cost-effective way of transferring technology (Frame, 1983).

A Case Study

The process of transferring technologies through employee training is illustrated through a case study of a multinational consulting firm (Andersen Consulting). The firm's activities provide a framework for understanding how multinational firms' corporate training facilitate international technology transfer.

The Organization

Andersen Consulting is the largest information-systems consulting firm of its kind in the world. With a worldwide revenue of $1,441 million in 1989, the company has 157 offices and 18,143 professional employees in 45 countries. More than 50% of Andersen's professionals are in overseas operations, and most of them are local citizens.

A unique characteristic of Andersen Consulting is that it has espoused the "one firm" concept around the world, which is translated into standardization and uniform implementation of its consulting methodology. In other words, Andersen consultants use the same methodology in their client engagements, whether they are in New York or Seoul. To put the concept into action, as early as 1940 Arthur Andersen (the founder of Arthur Andersen & Co.) instituted a centralized training program assuring that the firm's personnel, regardless of where they were located, underwent uniform training and met common standards of client service. In the fiscal year of 1989, Andersen Consulting spent $126 million in providing 140,000 training days. The heavy reinvestment in training, support tools, standardization, and uniform implementation of its consulting methodology has stimulated much international technology transfer (Noling & Blumental, 1985).

Technology transfer at Andersen Consulting occurs mainly through two sources—the world headquarters (WHQ) in Chicago, Illinois and the professional education center (PED) in St. Charles, Illinois. The transfer process can be diagramed by tracing the information flow within the firm and explaining the activities of the consultants in their client engagements.

The World Headquarters (WHQ)

Andersen Consulting world headquarters provides a strategic direction for the firm as well as developing the firm's products and standard methodologies that are marketed and used throughout the world. To ensure its clients will have the same service regardless of their locations, Andersen's world headquarters assists its local offices in accepting, understanding, and using the firm's technologies. Andersen Consulting's international operations are headed by managing partners of three geographic areas—the Americas, Europe/Middle East/Africa, and Asia/Pacific—and six divisions (see Figure 13.1).

Each geographical area is grouped into several regions, each of which has its own area managing partner. In addition, each office has a consulting managing director who is ultimately responsible for the delivery of services and products to clients within that office's geographical responsibility. The functions that are of direct significance for the transfer of technology relate to product development, training, and client service. More specifically, these are technology services, application products, and international service lines which work closely with line offices around the world to disseminate new product development or methodology.

Technology Services

The technology-services section at Andersen's WHQ plays a central role in transferring technologies to the worldwide organization and to its clients. It has 500 people in two divisions—practice services, and research and development.

The practice-services division develops and disseminates state-of-the-art technologies, techniques, and methodologies by assisting the worldwide organization in promoting and executing client work. The division provides information about new products and services and ensures the availability of specialized technical training and self-study to the worldwide organization and its clients. The practice services division is organized into groups, each of which specializes in a particular technology area—including systems integrations, telecommunications, and artificial intelligence.

The research and development division is organized into two main groups. One group introduces new technology to the worldwide organization and to its clients. The second group is responsible for the development and conduct of training. The training group supports the

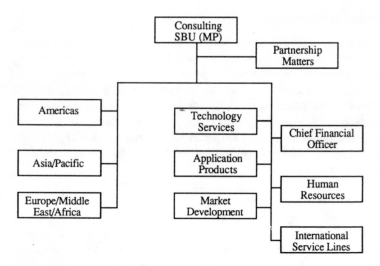

Figure 13.1. Andersen Consulting Worldwide Organization

*SBU stands for Strategic Business Unit and MP for Managing Partner.

firm's needs by providing knowledge transfer and training in key technical, functional, and industry areas of Andersen Consulting's practice. Priorities and directions for consulting training are set with the help of a management and advisory committee consisting of different practice and program partners.

On-the-Job Training

Another role which the WHQ plays in technology transfer is to provide on-the-job training to the employees from overseas. In 1989, about 60% of the people at the technology services section were from foreign offices working on long-term assignments (12 to 18 months). Some were engaged in developing the firm's new products or methodology; others were engaged in adapting existing products to new products that would meet local market needs such as translating the firm's standard product into local languages. On-the-job training at the WHQ not only helps employees develop new technical skills, but also provides the means to communicate corporate culture and to share values in global operations that are essential to coordinated business development.

Professional Education Center (PED)

The professional education center is primarily a centralized training facility for Andersen Consulting personnel, but it also helps clients develop their own industry, business, and technology skills. Located on 145 acres of wooded land in St. Charles, Illinois, it has accommodations for 1,700 people, 135 training and meeting rooms of various sizes, five conference centers, six auditoriums, two amphitheaters, and a social center. With instructional design and management development staff, it aims to provide education for both consultants and clients.

Formal Training at Andersen

Formal training for Andersen consultants typically starts at their respective offices. The first week is for the orientation program for new hires. During the orientation, they meet groups of people in different capacities who help new consultants understand the firm's philosophy, business objectives, and basic engagement methodologies. Recruits study the firm's products and engagement tools using paper-based materials and computer-based training (CBT). Each practice specialty requires different schools; however, the typical training period for new consultants is six weeks of their first two years of employment with Andersen.

Client Education

Andersen consultants often work with clients' project teams and end users so that both the process of system design and the system itself are as productive as possible. For example, in various engagements, Andersen Consulting provides user training and procedure documents. In the process, consultants in technical areas (e.g., systems integration) work with consultants in the change-management services area who specialize in knowledge transfer and technology assimilation. As a team, they design and develop technology-integrated user training and procedures for clients. When appropriate, cutting-edge technologies are used, such as interactive videodisc (IVD), digital video interactive (DVI), computer-based training (CBT) and video conferencing.

As the number of Andersen's clients has increased worldwide, the need for training has also increased. Andersen Consulting offers centralized, regularly scheduled courses to clients at its Center for Professional Development. The centralized training (which brings several clients together at one time) provides a forum for interaction and discussion. These seminars and training programs are critical because users not only must be able to understand the system's logic and functions, but they must also be able to develop appropriate operation policies and interpret information to help in their decision making process. Furthermore, client education can be a way of disseminating technology to a large group of people.

Because all instruction for centralized training is conducted in English, the ability to communicate in English becomes an important issue for successful technology transfers. Although most consultants in Andersen's foreign offices can understand English, only those who have a good command of English can make the best use of their training.

Personnel Exchange

Personnel exchange is another transfer mechanism. Often Andersen Consulting has a multinational team for projects in a given country. In 1988, there were about 50 Americans on loan to Andersen's United Kingdom consulting practice. The UK clients also had consulting personnel from Australia, Canada, Denmark, Ireland, Malaysia, Portugal, Singapore, Spain, and Sweden. This team approach not only expands the capabilities of staff consultants, but also helps them learn to work cooperatively with internationals.

Personnel exchange within a region is most common. A project in Singapore is likely to be staffed with consultants from Korea (rather than someone from France) if there is a need for additional personnel. In some cases, using consultants from a less developed country for a project in a developed country may serve two purposes: (1) to benefit the consultants and their offices by expanding technical capabilities and information networks from the engagement, and (2) perhaps to offer clients a competitive fee because their salaries are lower.

Summary and Discussion for Case Study

Not all technical information exchange at Andersen Consulting is recorded; however, it is clear that technology is being transferred to many countries through communication between Andersen world headquarters and overseas offices and in turn to their clients. Figure 13.2 presents the general picture of information exchange at Andersen Consulting which facilitates technology transfer directly and indirectly. WHQ transfers knowledge and skills necessary to use products and methodologies to line office personnel (who serve clients directly)

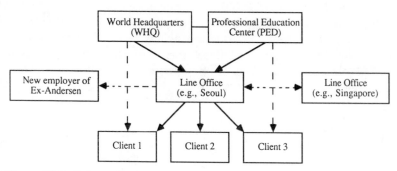

Figure 13.2. Information Flow at Andersen Consulting

through on-the-job training at the WHQ or centralized training at PED. In turn, the line office personnel transfer their knowledge and skills to clients through working in a team with clients or providing user seminars and documenting user procedures. In some cases, the WHQ transfers technologies directly to clients by sending their experts to client sites. The Professional Education Center also may transfer Andersen's technologies directly to clients through its client education programs.

In addition, whenever an Andersen consultant leaves the firm, that individual takes the technical skills learned from Andersen's practice. In this case, technology is transferred from Andersen Consulting to a new employer in the local market through a certain employee. Although turnover rates at local Andersen offices are unavailable, personnel turnover seems to be an unintended mechanism of technology transfer.

Although the current centralized training has been the main vehicle through which Andersen Consulting has accomplished the task of providing its consultants training and development, with the rapid increase in the number of overseas consulting professionals, the high cost of running centralized training programs has begun to be recognized as a problem. Recently, the first school for consultants in the systems-integration field has changed its approach from using mainframes to using personal computers (PCs) in teaching programming concepts and providing hands-on programming practices. This change will save Andersen Consulting more than $30 million over the next five years, because the costs of supporting a PC are much lower than support costs on a mainframe. Using PCs will also require fewer faculty because the PC allows for greater self-instruction. Eventually, it will enable some classes to be held at local offices, rather than at the Professional Education Center. The new approach may be as effective as the old one in teaching the technical skills to consultants; however, it means a loss of control and a forum for technology transfer at which consultants from different countries meet and develop networks for future information exchange.

Conclusion

The challenge to successful international technology transfer lies in achieving optimal transactions between donors and recipients. Some pessimism exists regarding whether international technology transfer will ever occur to the satisfaction of all parties; however, strategic alliances between donors (multinational firms) and recipients (less developed countries) are critical to continuing technology transfer. In the future, developing countries are likely to demand more transfer of the capability to develop their own technologies, rather than simply transfer of ready-made products. Consequently, multinational training and education for local employees and clients will be welcomed as an opportunity to strengthen local capabilities.

Since instructional technologies using telecommunications travel long distances, more companies will be likely to replace their central training with less costly media-based education. Before adopting technology-delivered courses, however, multinationals should be reminded that technology-based courses are only suitable for certain subjects and target groups. Technology-based training also will lead to a loss of opportunity for employees from different offices and countries to exchange ideas. In future studies, it will be interesting to investigate how decentralized training using high-tech instructional technologies will affect technology transfer as compared to more centralized training.

References

Carnevale, A. (1989, February). The learning enterprise. *The Training and Development Journal*, 26-33.

Eurich, N. P. (1985). *Corporate classroom: The learning business.* New York: Carnegie Foundation for the Advancement of Teaching.

Ezzat, H. A., Howell, L. J., & Kamal, M. M. (1989). Transferring technology at General Motors. *Research-Technology Management, 32,* 32-35.

Frame, J. D. (1983). *International business and global technology.* Lexington, MA: Lexington.

Galagan, P. A. (1989, January). IBM gets its arms around education. *Training and Development Journal,* 34-40.

Goldstein, M. H. (1988). Office automation: Technology training. *National Productivity Review, 7,* 73-76.

Malaysia: Demand for management and technical consulting services expected to grow. (1989). *East Asian Executive Reports, 11,* 8, 21, 23.

Nees, D. B. (1986). Building an international practice. *Sloan Management Review, 27,* 15-26.

Noling, M. S., & Blumental, J. F. (1985). Gaining competitive advantage as a professional service firm. *New Management, 3,* 53-57.

Ohmahae, K. (1989, March-April). The global logic of strategic alliance. *Harvard Business Review,* 143-154.

Robinson, R. D. (1988). *The international transfer of technology.* Cambridge, MA: Ballinger.

Stiffler, A. (1985, October). Management, consulting services continues rapid overseas growth. *Business America, 8,* 13-14.

Strizich, M. (1988, May). Information consulting: the tools of the trade. *On-Line, 12,* 27-31.

V

The Literature of Technology Transfer

14

The State of the Field:
A Bibliographic View of
Technology Transfer

DAVID V. GIBSON
FREDERICK WILLIAMS
KATHY L. WOHLERT

This chapter reports on a preliminary search and analysis of the literature on technology transfer and technological innovation conducted by Kathy L. Wohlert.[1] Technology transfer is examined along four key dimensions: content, context, consequences, and communication. The content is the technological innovation itself. The context is the situational environment in which the transfer occurs, including its geographical, cultural, and political milieu. The third dimension, consequences, describes the legal, ethical, and economic aspects of the transfer phenomenon. The final dimension, communication, constitutes the social infrastructure for technology transfer. Perhaps most important to understanding this dimension is the recognition that organizations do

not communicate—people do. Gibson and Williams' backgrounds
are described in the introductory chapter of this volume. Kathy L.
Wohlert is a doctoral student in Organizational Communication
at the University of Texas at Austin.

Developing the Bibliography

Five data bases were selected for the preliminary search of the
technology transfer literature: (1) ABI/Inform (a business data base),
(2) Management Contents, (3) Dissertation Abstracts International, (4)
PAIS (Public Affairs Information System), and (5) InfoTrac (business,
social science, humanities, and general interest periodicals from 1985
to the present). Data bases contain only journal or periodical listings.
Additional sources from which the initial set of bibliographic data were
derived included books, journal articles, conference papers, government hearings, unpublished doctoral dissertations and papers, and draft
manuscripts of forthcoming books. The original set of data consisted of
850 citations, with more than 75% containing the key words *technology
transfer* in the title. We have not included references and citations that
appear in the bibliographies of the preceding chapters.

Some General Characteristics of the Literature

Content Areas

In the initial analysis of the first three data bases (ABI, Management
Contents, and DAI), each entry was categorized as either "international" or "United States" based on the title. Using this classification,
40% of the articles found in the ABI data base were international. For
the Management Contents data base, 46% were international. And for
the Dissertation Abstracts International data base entries, 70% of the
sources were internationally focused. All but two of the dissertations
were written by students attending universities in the United States.
Clearly, one of the distinguishing characteristics of this literature is the
preponderance of information on technology transfer as an international
phenomenon. Also noteworthy was the virtual absence of articles with
a communication perspective on technology transfer. Furthermore, only

three of the 450-plus titles examined referred to the educational or training aspects of technology transfer. It is assumed, however, that a search of the educational data base ERIC would be fruitful in this area.

The Public Affairs Information System data base was evaluated to determine its primary emphasis. Of the 493 English articles referenced in this data base using the key words *technology transfer*, more than 90% focused on international transfer, LDCs (lesser developed countries), regulations, ethics, economics, and finance. Again, neither communication nor education rated a singular entry.

Two separate searches were conducted on the InfoTrac data base using the key words *technology transfer* and *technological innovation* to determine if any differences existed in the primary orientation of the literature. The profile provided by the data from the InfoTrac data base (Table 14.1) indicates that transfer and innovation overlap in many subject matter areas, but they can be differentiated on the basis of the *amount* of attention devoted to them in the literature.

The topics in bold print illustrate the contrast effect. For example, under the topic of developing countries, 28 articles are cited in the transfer literature and only 6 are cited in the innovation literature. Even more dramatic contrasts exist in areas associated with international, policy, legal, and security issues, with the literature being weighted heavily in favor of technology transfer. That is, there are at least three technology transfer articles for every technological-innovation article in these areas of study.

On the other hand, technological innovation articles enjoy a 3:1 ratio to technology transfer articles in the following areas: business applications, economic aspects, finance, forecasts, management, and psychosocial aspects. Both topics share almost equal space in the literature on topics related to analysis, planning, and Japan.

Journal Sources

From the almost 800 journal entries examined, 24 periodicals appeared seven or more times. Table 14.2 provides a summary of the names of these journals and the frequency of citations. Seven of the 24 journals originate in the United Kingdom, the Netherlands, Switzerland, or Canada. Indeed, technology transfer is not a U.S. monopoly; utilizing these international communication sources may well facilitate more of a global understanding of technology transfer.

TABLE 14.1: Profile of INFO TRAC (Business, Social Science, Humanities, and General Interest Periodicals, 1985-Present)

TECHNOLOGY TRANSFER (TT)

Other key words referenced included:
Foreign licensing agreements
   Information technology
   Nuclear nonproliferation
   Technological forecasting
   Technological innovations
   Technology-international cooperation

TECHNOLOGICAL INNOVATION (TI)

Other key words referenced included:
   Agricultural innovation
   Diffusion of innovation
   Education, Higher—effects of technological innovations
   Employees—effects of technological innovations
   Forestry innovations
   Industrial productivity—effects of tech innovations
   Labor supply—effects of technological innovations
   Literacy—effects of technological innovations
   Machinery in industry
   Medical innovation
   Office workers—effects of technological innovations
   Research, industrial
   Technology assessment
   Technology transfer

TOPIC	TT	TI	TOPIC	TT	TI
accounting	—	1	bars, saloons	—	1
achievement/awards	—	8	books	—	2
addresses, essays	—	5	**business applications**	**4**	**30**
aeronautic use	—	1	California	2	3
Africa	1	2	Canada	2	3
aims & objectives	1	1	case studies	7	5
analysis	**48**	**50**	**cases**	**26**	**1**
anecdotes, etc.	—	6	charts, tables	1	1
appreciation	—	2	**China**	**29**	**2**
Arizona	—	1	citizen participation	1	1
Asia	—	1	communication systems	1	1
Asia, Southeast	1	—	comparative methods	1	—
audio-visual arts	1	1	competitions	—	1
auditing/inspecting	1	1	contracts/specs	7	2
automobile	**—**	**10**	control	1	—

Table 14.1 continued

TOPIC	TT	TI	TOPIC	TT	TI
costs	—	2	information services	3	—
crime	5	1	innovations	1	—
Czechoslovakia	1	—	**international aspects**	97	30
data bases	1	—	**international cooperation**	43	12
decision making	2	3	**investigations**	19	—
demographic aspects	—	1	investment activities	—	1
developing countries	28	6	Iowa	1	—
economic aspects	30	153	**Japan**	21	17
educational use	1	4	joint ventures	4	2
electric industries	1	1	Korea, Republic of	1	—
employee participation	—	12	Korea, South	3	1
employment	—	1	Latin America	4	—
energy consumption	—	1	**law & legislation**	23	3
environmental aspects	1	1	legal aspects	3	3
equipment/supplies	—	2	library applications	—	3
ethical aspects	6	2	licenses	2	2
Europe	1	2	litigation	3	—
Europe, Western	2	—	Louisiana	1	—
evaluation	1	4	**management**	21	104
exhibitions	—	7	manufacture	—	1
export/import problems	—	1	manufactures	—	3
export-import trade	2	3	**marketing**	4	27
federal aid	1	2	markets	—	4
fiction	—	1	**math models**	8	29
finance	1	15	measurement	—	2
forecasts	6	45	mediation & arbitration	2	—
foreign influences	—	1	medical care	—	7
France	—	1	medical use	—	3
galleries/museums	—	1	methodology	4	1
government policy	55	10	Michigan	1	—
grants	1	—	Micronesia	—	1
Great Britain	1	4	Middle East	1	—
growth/development	—	1	**military aspects**	20	8
history	1	22	**military use**	—	10
home use	—	1	Minnesota	—	1
Hungary	1	—	models	—	1
Illinois	1	1	moral and		
indexes	—	1	religious aspects	—	2
India	5	2	moving picture use	—	1
Indiana	1	—	musical use	—	1
industrial applications	3	27	Nigeria	—	1
industries	—	1	Ohio	2	2
influence	—	7	Oklahoma	—	1

(continued)

Table 14.1 continued

TOPIC	TT	TI	TOPIC	TT	TI
Oregon	—	1	South America	—	1
origin	—	1	**Soviet Union**	39	11
Pakistan	3	3	Spain	1	—
patents	—	7	sports use	—	1
Pennsylvania	1	4	standards	—	1
periodicals	—	1	statistics	—	1
personal narratives	—	2	**steel industry**	—	10
personnel management	—	12	strategic aspects	1	—
planning	17	20	study & teaching	—	1
Poland	1	1	surveys	4	5
political aspects	39	7	Sweden	—	1
prevention	15	—	Taiwan	—	1
product development	—	4	taxation	1	—
product discontinuation	—	1	technique	3	—
production control	—	1	technological innovation	1	—
production methods	—	5	temperature	—	1
productivity	—	1	Tennessee	—	1
protection	—	2	Texas	—	2
psychological aspects	1	16	training	—	1
public opinion	3	12	transportation	—	1
purchasing	—	1	United States	4	4
reports	12	2	urban areas	—	1
research	11	43	**usage**	—	12
rules/regulations	47	4	use studies	—	1
rural areas	1	—	user education	—	3
scientific applications	1	3	valuation	—	1
security measures	42	4	Yugoslavia	—	1
seminars, etc.	2	3	20th century	—	1
services	2	—	1900-1989	—	1
social aspects	7	61	1986	—	1
sociological aspects	1	22	1987	—	1
sources	—	2	1988	—	4
South Africa	1	—			

Selections for the Bibliography

Two primary criteria were applied in the selection of entries for the following bibliography. The first was a social dimension criterion that focused on the individual or interpersonal level of technology transfer. That is, articles and books were selected if they reflected a communication or management perspective. Another group of citations were

TABLE 14.2: Frequency of Journal Citations

NAME OF JOURNAL	COUNTRY OF ORIGIN	FREQUENCY
Research Management	U.S.	49
R&D Management	United Kingdom	28
Academy of Management Journal	U.S.	27
Administrative Science Quarterly	U.S.	27
Technological Forecasting & Social Change	U.S.	24
International Journal of Technology Management	Switzerland	22
Research Policy	The Netherlands	19
IEEE Transactions on Engineering Management	U.S.	19
Society of Research Administrators Journal	U.S.	17
Science	U.S.	16
Technovation	United Kingdom	14
Columbia Journal of World Business	U.S.	11
Journal of Information Science	The Netherlands	9
Business Quarterly	Canada	9
Journal of World Trade Law	Switzerland	8
Academy of Management Review	U.S.	8
Vital Speeches	U.S.	8
American Sociological Review	U.S.	8
Business Week	U.S.	8
Human Relations	U.S.	7
Technology Review	U.S.	7
Harvard Business Review	U.S.	7
Journal of Industries Economics	U.S.	7

included because of their general orientation to technology transfer as an intra- or interorganizational phenomenon. These articles emphasize the concepts and linkages involved in studying technology transfer.

Notes

1. A comprehensive data base of more than 3,000 entries on technology transfer and technological innovation has been compiled. For more information, contact The Center for Research on Communication Technology and Society, College of Communication, The University of Texas at Austin, Texas 78712, (512) 471-5826.

2. Recently, Cottrill and Rogers (1990) contrasted the two processes of technology transfer and the diffusion of innovation in a co-citation analysis of the literature. Their

findings suggest that researchers studying technology transfer are affiliated predominantly with the disciplines of economics, business management, and engineering, whereas diffusion-of-innovation scholars are identified with the disciplines of sociology and communication.

Bibliography

Abelson, P. (1986). Evolving state-university-industry relations. *Science, 231.*

Aldrich, H., & Herker, D. (1977). Boundary spanning roles and organizational structure. *Academy of Management Review, 2,* 217-230.

Ali, S. N. (1989). Science and technology information transfer in developing countries: Some problems and suggestions. *Journal of Information Science Principles and Practice, 15,* 82-93.

Allen, T. J. (1966). Performance information channels in the transfer of technology. *Industrial Management Review, 8,* 87-98.

Allen, T. J., Tushman, M. L., & Lee, D. M. S. (1979). Technology transfer as a function of position in the spectrum from research through development to technical services. *Academy of Management Journal, 22*(4), 694-708.

Anderson, P. V. (1984). What technical and scientific communicators do: A comprehensive model for developing academic programs. *IEEE Transactions on Professional Communication, PC-27*(3), 161-167.

Ansoff, H. I. (1987). Strategic management of technology. *Journal of Business Strategy, 7,* 28-40.

Ardisson, J. M., & Bidault, F. (1986). Technology transfer strategies as a means to build international networks. *Industrial Marketing & Purchasing, 1*(1), 59-71.

Armes, G. (1984). The key to technology transfer. *Satellite Communications, 8,* 27-30.

Avery, C. (1989). *Organizational communication and technology transfer between an R&D consortium and its shareholders.* Unpublished doctoral dissertation, College of Communication, University of Texas at Austin.

Axley, S. R. (1984). Managerial and organizational communication in terms of the conduit metaphor. *Academy of Management Review, 9*(3), 428-437.

Azaroff, L. (1982). Industry-university collaboration: How to make it work. *Research Management, 25,* 31-34.

Badawy, M. K. (1989). Integration: The fire under technology transfer. *Industry Week, 238,* 39-77.

Baldwin, D. R., & Green, J. W. (1984). University-industry relations: A review of the literature. *Journal of the Society of Research Administrators, XV*(4).

Barber, A. A. (1985). University-industry research cooperation. *Journal of the Society of Research Administrators, 17,* 19-31.

Baroco, P. E. (1983). Technology transfer: What's acceptable and why—industry's view. *Security Management, 27,* 51-56.

Baty, G., Evan, W., & Rothermel, T. (1971). Personnel flows as interorganizational relations. *Administrative Science Quarterly, 16,* 430-443.

Bitting, R. K. (1988). Observations from Japan: Lessons in research and technology transfer. *Journal of the Society of Research Administrations, 19*(4), 17-22.

Bremer, H. W. (1985). Research applications and technology transfer. *Journal of the Society of Research Administrators, 17*, 53-65.

Bright, J. R. (1986, February). Improving the industrial anticipation of current scientific activity. *Technological Forecasting & Social Change*, 1-13.

Brown, W. B. (1966). Systems, boundaries, and information. *Academy of Management, 9*, 318-327.

Brown, W. B., & Schwab, R. C. (1984). Boundary-spanning activities in electronics firms. *IEEE Transactions on Engineering Management, EM-31*, 105-111.

Bullock, M. (1985). Cohabitation: Small research-based companies and the universities. *Technovation, 3*, 27-38.

Carroad, P. A., & Carroad, C. A. (1982). Strategic interfacing of R&D and marketing. *Research Management, 25*, 28-33.

Carter, N. E. (1986). The political side of science: Communication between scientists and the public. *Vital Speeches, 52*(18), 558-561.

Chakrabarti, A. K., Feinman, S., & Fuentevilla, W. (1982). Targeting technical information to organizational positions. *Industrial Marketing Management, 25*, 28-33.

Cheng, J. L. C. (1984). Paradigm development and communication in scientific settings: A contingency analysis. *Academy of Management Medical, 27*(4), 870-877.

Cohen, H., Keller, S., & Streeter, D. (1979). The transfer of technology from research to development. *Research Management, 22*, 11-17.

Cook, J. (1985, December 16). It's just too powerful a force. *Forbes, 118*, 120.

Corsten, H. (1987). Technology transfer from universities to small- and medium-sized enterprises: An empirical survey from the standpoint of such enterprises. *Technovation, 6*(1), 57-68.

Cottrill, D., & Rogers, E. M. (1990). *A co-citation analysis of the research traditions of technology transfer and diffusion of innovation.* Unpublished manuscript, Annenberg School for Communication, University of Southern California, Los Angeles.

Creighton, J. W., Jolly, J. A., & Buckles, T. A. (1985). The manager's role in technology transfer. *Journal of Technology Transfer, 10*, 67.

Cutler, R. S. (1989). A comparison on Japanese and U.S. high-technology transfer practices. *IEEE Transactions on Engineering Management, 36*, 17-24.

Cyert, R. M. (1985). Establishing university-industry joint ventures. *Research Management, 28*, 27-28.

Dahlman, C., & Westphal, L. (1983). The transfer of technology. *Finance & Development, 20*, 6-9.

Das, S. (1987). Externalities and technology transfer through multinational corporations: A theoretical analysis. *Journal of International Economics, 22*, 171-182.

Davidson, W. H., & McFetridge, D. G. (1985). Key characteristics in the choice of international technology transfer. *Journal of International Business Studies, 16*, 5-21.

Davies, H. (1977). Technology transfer through commercial transactions. *Journal of Industrial Economics, 26*(2), 161-175.

Davis, B. G., & Simpson, J. K. (1987). A simple and economical approach to developing a technology transfer program. *Journal of the Society of Research Administrators, 19*, 15-19.

Davis, P., & Wilkof, M. (1988). Scientific and technical information transfer for high technology: Keeping the figure in its ground. *R&D Management, 18*, 45-58.

Dean, C. W. (1981). A study of university/small business for interaction for technology transfer. *Technovation, 1*(2), 109-123.

De Fleur, M. L. (1988). Diffusing information. *Society, 25,* 72-82.

Dembo, V. (1979). Technology transfer planning. *R&D Management, 9,* 117-124.

Deutsch, M. (1949). A theory of cooperation and competition. *Human Relations, 2,* 129-152.

Devine, M. D., James, T. E., Jr., & Adams, T. I. (1987). Government supported industry-university research centers: Issues for successful technology transfer. *Journal of Technology Transfer, 12,* 27-37.

Dollinger, M. J. (1984). Environmental boundary spanning and information processing effects on organizational performance. *Academy of Management Journal, 27,* 351-368.

Dorf, R. C., & Worthington, K. K. F. (1989). Technology transfer: Research to commercial product. *Engineering Management International, 5*(3), 185-191.

Doz, Y. L. (1987). Technology partnerships between larger and smaller firms: Some critical issues. *International Studies of Management and Organization, 17,* 31-58.

Fischer, W. A., Zmud, R. W., Hamilton, W., & McLaughlin, C. P. (1986). The elusive product champion. *Research Management, 29,* 13-17.

Friedkin, N. E. (1978). University social structure and social network among scientists. *American Journal of Sociology, 83,* 1444-1465.

French, E. B. (1966). Perspective: The motivation of scientists and engineers. *Academy of Management Journal, 9*(3), 152-156.

Frosch, R. A. (1984). Linking R&D with business needs: R&D choices and technology transfer. *Research Management, 27,* 11-14.

Fusfield, H. I., & Haklisch, C. S. (1985). Cooperative R&D for competitors. *Harvard Business Review, 63,* 60-76.

Galaskiewicz, J. (1985). Interorganizational relations. In R. Turner (Ed.), *Annual Review of Sociology,* (pp. 281-304). Palo Alto, CA: Annual Reviews.

Gander, J. P. (1987). University-industry research linkages and knowledge transfers: A general equilibrium approach. *Technological Forecasting & Social Change, 31.*

Gartner, J., & Naiman, C. S. (1978). Making technology transfer happen. *Research Management, 21,* 34-38.

Gee, S. (1981). *Technology transfer, innovation, and international competitiveness.* New York: Wiley.

Gerstenfeld, A., & Berger, P. (1980). An analysis of utilization differences for scientific and technical information. *Management Science, 26*(2), 165-179.

Gervais, B. (1988). Planning for technology transfer. *International Journal of Technology Management, 3,* 217-224.

Gibson, W. D. (1985, June 12). Tracking technology around the globe. *Chemical Week,* 34-35.

Gilchrist, A. (1986). What the information scientist has to offer. *Journal of Information Science Principles and Practice, 12*(6), 273-281.

Ginn, M. E., & Rubenstein, A. H. (1986). The R&D/production interface: A case study of new product commercialization. *Journal of Product Innovation Management, 3*(3).

Godkin, L. (1988). Problems and practicalities of technology transfer: A survey of the literature. *International Journal of Technology Management, 3*(5), 587-603.

Goldhor, R. S., & Lund, R. T. (1983). University-to-industry advanced technology transfer: A case study. *Research Policy, 12*(3), 121-152.

Goldstein, M. L. (1987, November 16). Gatekeepers: Fit the key into new technology. *Industry Week,* 26-36.

Goldstone, N. J. (1986). How not to promote technology. *Technology Review, 89*, 22-23.

Gore, A. P., & Lavaraj, U. A. (1987, September). Innovation diffusion in a heterogeneous population. *Technological Forecasting & Social Change*, 163-168.

Gray, B. (1985). Conditions facilitating interorganizational collaboration. *Human Relations, 38*, 911-936.

Gray, P. E. (1982). Technology transfer at issue: The academic viewpoint. *IEEE Spectrum, 19*, 64-68.

Guterl, F. V. (1987). The dilemma of technology transfer. *Business Month, 41*.

Guterl, F. V., & Hollyday, A. (1987, September). Technology transfer isn't working. *Business Month*, 44-48.

Hall, R. H., Clark, J. P., Giordano, P. C., et al. (1977). Patterns of interorganizational relationships. *Administrative Science Quarterly, 22*, 457-472.

Hammer, D. (1988, April). Let the buyer beware: Technology rights are key to takeovers. *High Technology Business*, 17.

Haughey, C. S. (1979). Technology transfer in a competitive environment. *Defense Systems Management, 2*, 27-34.

Hill, J. S., & Still, R. R. (1980). Cultural effects of technology transfer by multinational corporations in lesser developed countries. *Columbia Journal of World Business, 15*, 40-51.

Hirschey, R. C., & Caves, R. E. (1981). Research and transfer of technology by multinational enterprises. *Oxford Bulletin of Economics and Statistics, 43*(2), 115-130.

Huber, G. H., & Daft, R. L. (1987). Information environments. In F. Jablin, L. Putnam, K. Robert, & L. Porter, (Eds.), *Handbook of organizational communication*. Newbury Park, CA: Sage.

Irving, R. R. (1983). Technology transfer: That's what it's all about. *Iron Age, 226*, 53-62.

Jacobson, C. (1982). The technology transfer issue. *Business America, 5*, 2-5.

Jarillo, J. C. (1988). On strategic networks. *Strategic Management Journal, 9*, 31-42.

Jeannet, J. P., & Liander, B. (1978). Some patterns in the transfer of technology within multinational corporations. *Journal of International Business Studies, 9*, 108-118.

Kedia, B. L., & Bhagat, R. S. (1988). Cultural constraints on transfer of technology across nations: Implications for research in international and comparative management. *Academy of Management Review, 13*, 559-571.

Keller, R., & Holland, W. (1975). Boundary spanning roles in a research and development organization: An empirical investigation. *Academy of Management Journal, 18*, 388-393.

Kennedy-Minott, R. (1983). Technology transfer: An overview. *Telecommunications, 17*, 104-108.

Klimstra, P. D. (1988). What we've learned: Managing R&D projects. *Research-Technology Management, 31*, 23-39.

Knight, K. E. (1967). A descriptive model of the intra-firm innovation process. *The Journal of Business, 40*, 478-496.

Kozmetsky, G. (1989). Tomorrow's transformational managers. In K. D. Walters, (Ed.), *Entrepreneurial management: New technology and new market development*. Cambridge, MA: Ballinger.

Larson, J. K., Wigand, R. T., & Rogers, E. M. (1986). *Industry-university technology transfer in microelectronics*. Los Altos, CA: Cognos and Associates.

Lasserre, P. (1982). Training: The key to technological transfer. *Long Range Planning, 15*, 51-60.

Lasserre, P. (1984). Selecting a foreign partner for technology transfer. *Long Range Planning, 17,* 43-49.

Le Coadic, Y. F. (1987). Modelling the communication, distribution, transmission or transfer of scientific information. *Journal of Science Principles and Practice, 13*(3), 143-148.

Ledbetter, J. (1987). Technology transfer. *Computerworld,* 63-68.

Leifer, R., & Huber, G. (1977). Relations among perceived environmental uncertainty, organizational structure, and boundary spanning behavior. *Administrative Science Quarterly, 22,* 235-247.

Leonard-Barton, D. (1985). Experts as negative opinion leaders in the diffusion of a technological innovation. *Journal of Consumer Research, 11,* 914-927.

Levine, S., & White, P. E. (1961). Exchange as a conceptual framework for the study of interorganizational relationships. *Administrative Science Quarterly, 5,* 583-601.

Lindgren, N. A. (1987, April 2). Consortium-style research and development proves its worth. *Public Utilities Fortnightly,* 26-33.

Machlup, F. (1962). *The production and distribution of knowledge in the U.S.* Princeton, NJ: Princeton University Press.

Machlup, F. (1980). *Knowledge: It's creation, distribution, and economic significance, Volumes 1-8.* Princeton, NJ: Princeton University Press.

Madu, C. N., & Jacob, R. (1989). Strategic planning in technology transfer: A dialectic approach. *Technological Forecasting and Social Change, 35*(4), 327-338.

Mansfield, E., & Romeo, A. (1984). 'Reverse' transfer of technology from overseas subsidiaries to American firms. *IEEE Transactions on Engineering Management, 31,* 122-127.

Marton, K. (1986). Technology transfer to developing countries via multinationals. *World Economy, 9*(4), 409-425.

McCardle, K. F. (1985). Information acquisition and the adoption of new technology. *Management Science, 31,* 1372-1390.

McClenahen, J. S. (1987, August 24). Alliances for competitive advantage. *Industry Week,* 33-36.

McDonald, D. W., & Gieser, S. M. (1987). Making cooperative relationships work. *Research Management, 30,* 38-42.

McPherson, J. (1985). Linking good minds together can spur ideas and bring results. *International Management, 40,* 59-60.

Millman, A. F. (1983). Technology transfer in the international market. *European Journal of Marketing, 17*(1), 26-47.

Mindlin, S. E., & Aldrich, H. (1975). Interorganizational dependence: A review of the concept and a reexamination of the findings of the Aston group. *Administrative Science Quarterly, 20,* 382-392.

Moravcsik, M. J. (1983). The role of science in technology transfer. *Research Policy, 12*(5), 287-296.

More, R. A. (1985). An overview of management problems and research needs. *Business Quarterly, 50,* 78-83.

Moritani, M. (1987). Technology transfer management. *Management Japan, 20,* 18-24.

Morone, J., & Alben, R. (1984). Matching R&D to business needs. *Research Management, 27,* 33-39.

Morone, J., & Ivins, R. (1981). Problems and opportunities in technology transfer from the national laboratories to industry. *Research Management, 25,* 35-44.

Myers, L. A. (1983). Information systems in research and development: The technological gatekeeper reconsidered. *R&D Management, 13,* 199-206.

O'Keefe, T. G., & Marx, H. (1986). An applied technology transfer process. *Journal of Technology Transfer, 11,* 83-88.

Oliver, D. S. (1982). Some aspects of technology transfer. *Journal of the Society of Research Administrators, 14,* 543-544.

Organ, D. W. (1971). Linking pins between organizations and environments. *Business Horizons, 14,* 73-80.

Organ, D. W., & Greene, C. H. (1972). The boundary relevance of the project manager's job: Findings and implications for R&D management. *R&D Management, 3,* 7-11.

Ouchi, W. G. (1980). Markets, bureaucracies, and clans. *Administrative Science Quarterly, 24,* 129-141.

Ounjian, M. L., & Carne, E. B. (1987). A study of the factors which affect technology transfer in a multilocation multibusiness unit corporation. *IEEE Transactions on Engineering Management, 34*(3), 194-201.

Pacey, A. (1983). *The culture of technology.* Cambridge: MIT Press.

Pelc, K. I. (1986). Management of R&D for reduction of technological delay: A strategic viewpoint. *R&D Management, 16,* 97-103.

Pettigrew, A. M. (1972). Information control as a power resource. *Sociology, 6,* 187-204.

Pfaffenberger, B. (1986). Research networks, scientific communication, and the personal computer. *IEEE Transactions on Professional Communication, PC 29,* 30-33.

Pfeffer, J., & Nowack, P. (1976). Joint ventures and interorganizational dependence. *Administrative Science Quarterly, 21,* 398-418.

Pinchot, G., III. (1987). Innovation through entrepreneuring. *Research Management, 30,* 14-19.

Posner, B. G. (1985, June). Strategic alliances. *Inc.,* 74-80.

Price, F. O. (1980). Technology transfer internationally: Why governments are concerned. *Journal of the Society of Research Administrators, 11,* 5-17.

Provan, K. G. (1982). Interorganizational linkages and influence over decision making. *Academy of Management Journal, 25,* 443-451.

Provan, K. G., Beyer, J. M., & Kruybosch, C. (1980). Environmental linkages and power in resource dependence relations between organizations. *Administrative Science Quarterly, 25,* 200-225.

Putnam, L. L., & Sorenson, R. L. (1982). Equivocal messages in organizations. *Human Communication Research, 8*(2), 114-132.

Rabino, S. (1989). High-technology firms and factors influencing transfer of R&D facilities. *Business Research, 18*(3), 195-205.

Rahm, D., Bozeman, B., & Crow, M. (1988). Domestic technology transfer and competitiveness: An empirical assessment of roles of university and governmental R&D laboratories. *Public Administration Review, 48*(6), 969-978.

Reekie, W. D., Allen, D. E., & Crook, J. N. (1984). On technological change, transfer and business characteristics: Some inferences from 12 case studies emanating from the National Engineering Lab. *Technovation, 2*(4), 233-254.

Reilly, D. J. (1988). Technology transfer: Successful only if process is managed. *Manufacturing Systems, 6*, 62-64.

Reynolds, L. (1989). Speeding transfer of technology. *Management Review, 78*, 56-58.

Robertson, T. S., & Gatignon, H. (1986). Competitive effects on technology diffusion. *Journal of Marketing, 50*, 1-14.

Rogers, D. M. A. (1989). Entrepreneurial approaches to accelerate technology commercialization. In K. D. Walters, (Eds.), *Entrepreneurial management: New technology and new market development,* (pp. 3-15). Cambridge, MA: Ballinger.

Rogers, D. M. A., & Dimancescu, D. (1987, April). *Managing the knowledge assets into the 21st century: Focus on research consortia.* Proceedings from a conference at Purdue University.

Rogers, E. M., & Bhowmik, D. K. (1970). Homophily-heterophily: Relational concepts for communication research. *Public Opinion Quarterly, 34*, 523-538.

Rogers, E. M., & Shoemaker, F. F. (1971). *Communication of innovations: A cross cultural approach.* New York: Free Press.

Sahal, D. (1982). *The transfer and utilization of technical knowledge.* Lexington, MA: D.C. Heath.

Scanlan, T., & Tushman, M. L. (1981). Boundary spanning individuals: Their role in information transmission and their antecedents. *Academy of Management Journal, 24*, 289-305.

Schermerhorn, J. R. (1975). Determinants of interorganizational cooperation. *Academy of Management Journal, 18*, 846-856.

Schermerhorn, J. R., Jr. (1977). Information sharing as an interorganizational activity. *Academy of Management Journal, 20*, 148-153.

Schwab, R. C., Ungson, G. R., & Brown, W. B. (1985). Redefining the boundary spanning-environment relationship. *Journal of Management, 11*, 75-86.

Schwartz, D. F., & Jacobson, E. (1977). Organizational communication network analysis: The liaison communication role. *Organizational Behavior and Human Performance,* 158-174.

Sethi, N. K., Movsesian, B., & Hickey, K. D. (1985, August). Can technology be managed strategically? *Long Range Planning,* 99-100.

Skowronski, S. (1987). Transfer of technology and industrial cooperation. *Technovation, 7*, 17-22.

Smilor, R. W., & Gibson, D. (1989). *Technology transfer at MCC: Findings and recommendations.* A research report for the MCC, IC^2 Institute, University of Texas at Austin.

Souder, W. E. (1987). The strategic management of technological innovations: A review and a model. *Journal of Management Studies, 24*, 25-42.

Spielman, J. D. B. (1983). Redefining the transfer of technology process. *Training and Development Journal, 37*, 35-40.

Spiker, B. K., & Daniels, T. D. (1981). Information adequacy and communication relationships: An empirical exam of 18 organizations. *Western Journal of Speech Communication, 45*(4), 342-354.

Stewart, C. T., Jr. (1987). Technology transfer vs. diffusion: A conceptual clarification. *Journal of Technology Transfer, 12*, 72.

Stubbart, C. (1982). Are environmental scanning units effective? *Long Range Planning, 15*, 139-145.

Sullo, P., Triscari, T., Jr., & Wallace, W. A. (1985). Reliability of communication flow in R&D organizations. *IEEE Transactions on Engineering Management, EM-32*(2), 91-97.

Tornatzky, L. G., Fergus, E. O., Avellar, J. W., & Fairweather, G. W. (1980). *Innovation and social process*. Elmsford, NY: Pergamon Press.

Tornatzky, L. G., & Fleischer, W. (1990). *The process of technological innovation*. Lexington, MA: Lexington.

Tsurumi, Y. (1979). Two models of corporation and international transfer of technology. *Columbia Journal of World Business, 14*, 43-50.

Tushman, M. L. (1977). Special boundary roles in the innovation process. *Administrative Science Quarterly, 22*, 587-605.

Tushman, M. L. (1978). Technical communication in R&D laboratories: The impact of project work characteristics. *Academy of Management Journal, 21*, 624-645.

Tushman, M. L., & Katz, R. (1980). External communication and project performance: An investigation into the role of gatekeepers. *Management Science, 26*, 1071-1085.

Tushman, M. L., & Scanlan, T. J. (1981a). Boundary spanning individuals: The role in information transfer and their antecedents. *Academy of Management Journal, 24*, 289-305.

Tushman, M. L., & Scanlan, T. J. (1981b). Characteristics and external orientations of boundary spanning individuals. *Academy of Management Journal, 24*, 883-898.

Ungar, E. W. (1986, March-April). America's united technologies: Using R&D results to meet the Japanese challenge—can consortia keep us competitive? *World*, 37-42.

Van de Bogart, W. (1981). Information management and technology transfer. *Information & Records Management, 15*, 26-28.

Van de Ven, A. H. (1976). On the nature, formation and maintenance of relations among organizations. *Academy of Management Review, 1*, 24-36.

Van de Ven, A. H., Emmett, D., & Koenig, R., Jr. (1974). Frameworks for interorganizational analysis. *Organization and Administrative Science Journal, 5*, 113-129.

Van de Ven, A. H., Walker, G., & Liston, J. (1979). Coordinating patterns within an interorganizational network. *Human Relations, 32*, 19-36.

Van Malderen, R. (1987). Worldwide telecommunications technology transfer. *International Journal of Technology Management, 2*(5), 649-660.

Wallich, P. (1982a). The dilemma of technology transfer. *IEEE Spectrum, 19*, 66-70.

Wallich, P. (1982b). Technology transfer: The industry viewpoint. *IEEE Spectrum, 19*, 69-73.

Webster, F. E., Jr. (1970). Informal communication in industrial markets. *Journal of Marketing Research, 7*, 186-189.

Whetten, D. A. (1981). Interorganizational relations: A review of the field. *Journal of Higher Education, 52*, 1-28.

Whetten, D. A., & Leung, T. K. (1979). The instrumental value of interorganizational relations: Antecedents and consequences of linkage formation. *Academy of Management Journal, 22*, 225-244.

White, W. (1977). Effective transfer of technology from research to development. *Research Management, 20*, 30-34.

Wolff, M. F. (1985). Bridging the R&D interface with manufacturing. *Research Management, 28*, 9-11.

Wren, D. A. (1967). Interface for interorganizational cooperation. *Academy of Management Journal, 10,* 69-81.

Zaltman, G., Dundan, R., & Holbeck, J. (1973). *Innovation and organizations.* New York: Wiley.

Zeitz, G. (1980). Interorganizational dialectics. *Administrative Science Quarterly, 25,* 72-88.

Index

Abernathy, W. J., 111, 129
Abetti, P. A., 159, 168
Ackerman, D., 110, 129
Adams, I. T., 15, 18
ADEPT, 55
Aerospace, 35, 136
Africa, 267
Agency for International Development (AID), 253-255
Agriculture, 110, 158, 163, 227, 229
Akers, L., 140
Alderfer, C. P., 98, 107
Aldrich, H., 67-69, 75-77, 80, 83, 85, 87
Alfred University, 180, 181
Allen, D. N., 164, 168
Alliance (strategic), 11, 12, 64, 85
Andersen Consulting, 266, 267, 269-275
Apple Computer, 180, 181
Applications, 14, 15, 17, 24, 28, 29, 37, 46, 94, 106, 115, 150, 158, 163, 175, 176, 180-182, 186, 207, 222, 231, 234-236, 241, 248, 267, 270
Argonne Tecnology Transfer Center, 155
Arizona, 17, 132
    Phoenix, 132, 134, 136, 139, 142
    Tempe, 139
Arizona State University, 17, 132, 135-139, 142
    Research Park, 138-141
Arms, C., 184, 190

ARPANET, 178
Arthur D. Little, Inc., 253, 256
Ashby E., 110, 129
Aslanian, C. B., 159, 170
Astley, W. G., 64, 65, 85, 87
AT&T, 74, 75, 94, 174, 183
Augustson, J. G., 190
Auster E. R., 17, 63, 64, 69, 70, 83, 113, 129
Austin Technology Incubator (ATI), 162
Australia, 273
Automobile Industry, 72, 73, 76-78, 204
Avery, C. M., 17, 93, 96, 107, 108

Baba M. L., 110, 129
Babbitt, B., 134
Backus, C. E., 136, 151
Baer, P., 182, 190
Balcones Research Park, 165-167
Baldwin, D. R., 122, 123, 130, 158, 169
Bangalore, 17, 240-249, 251-255
Barnes, J. A., 87
Barrera, E., 17, 195
BARRNet, 180, 181
Bauer, L. L., 112, 122, 123, 130
Beard, D., 84, 87
Beakley, G. C., 136, 151
Bell companies, 175
Bellcore, 95
Berlew, F., 64, 87

293

Berlo, D., 13, 18
Betz, M. F., 159, 170
Beyers, B., 112, 129
Biotechnology, 38, 65, 95, 156, 221, 244, 253
BITNET, 177
Bivens, K. K., 64, 87
Blau, P., 69, 87
Blevins, D. E., 133, 134, 152
Bloch, E., 133, 151
Blumenthal, J. F., 266
Boeing, 85
Boettcher, C., 137, 151
Boland, R. J., 49, 60
Bopp G., 156, 169
Boshwer, J., 264
Botkin, J., 15, 18, 95, 108, 159, 161, 169
Boundary spanners, 13, 80, 82, 119
Bozeman, B., 154, 170
Bozzo, U., 17, 226, 231, 238
BPI, 166, 167
Brackenridge, E., 17, 172
Bradbury, F., 44, 60
Brahm, R., 64, 65, 87
Bravo-Aguilera, L., 209
Brazil, 241
Brett, A., 118, 129, 163, 169
Brewster-Stearns, L., 80, 88
Bright, J. R., 108
Brookhaven National Lab, 180
Brown, F., 209
Brown, L. A., 44, 48, 60
Brown, T. L., 161, 169
Brown, W. S., 161, 169
Browning, L. D., 107, 108
Buckley, P., 69, 87
Burt, R., 68, 69, 76, 77, 87

California, 180, 181, 203
    San Francisco, 265
    San Jose, 141
Canada, 184, 273
Capital, 25, 28, 29, 39, 134, 155, 160, 162, 246-248, 254, 255
Carnevale, A., 262, 263, 272
Carrillo, J., 208, 209
Carroll, G., 85, 87
Caruana, C., 155, 169
Casson, M., 69, 87

Center for Developmennt of Telematics (C-DOT), India, 245, 254
Center for Professional Development, 136
Center for Technology Development (India), 253, 255
Center for Technology Venturing, 162
Centers for Advanced Teclnology (CATs), 182
Chen, Y. A., 241, 256
China, 38
Chmura, T., 151
Chrysler, 72, 73
Chu, P., 116
Collins T. C., 113, 123, 129, 158, 169
Columbia University, 17, 64, 180
Communication(s), 9-11, 13-15, 17, 48, 49, 53, 96, 98-100, 102-104, 106, 107, 113, 118, 132, 142, 145, 146, 150, 156, 172-175, 178, 180, 182, 189, 200, 201, 203, 205, 206, 212, 213, 215, 216, 218, 219, 222, 231, 233-236, 250, 252, 265, 269, 270
Community of Mediterranean Universities (CUM), 229
Competition/competitiveness, 11, 12, 22-26, 28, 31, 33, 35, 39, 64, 67, 81, 95, 105, 106, 110-112, 120, 133, 149, 153, 157, 159, 167, 180, 186, 187, 189, 207, 212, 262, 270
Computer, 52, 58, 94, 97, 156, 158, 172, 174-178, 187, 190, 200, 227, 235, 237, 241, 248, 251, 253, 255, 259, 265, 266
Computer Aided Design (CAD), 86, 98, 190, 237, 249
Consortium/consortia, 10-12, 17, 28, 29, 81, 85, 86, 93-96, 99, 100, 106, 113, 162, 174, 185, 189, 204, 229
Contractor, F., 64, 69, 87
Control Data Corporation, 94
Cook, K., 67, 87, 88
Cooke, R. A., 45, 62
Cooper, A. C., 156, 157, 169
Cooperation/cooperative, 12, 23, 30-32, 35, 39, 81, 93, 94, 103, 106, 110, 113, 117, 118, 120, 133, 134, 142-145, 149, 155, 158, 159, 164, 168, 178, 185, 187, 190, 199, 205, 212, 227-229, 233, 247, 248, 261, 262

Cornell University, 180, 181
Costa Rica, 265
Creativity, 22, 33-35, 178, 205
Crow, M., 154, 170
Crudele, J., 112, 130
Culnan, M. J., 49, 60
Culture, 10, 13, 26, 39, 53, 95, 103, 106,
    110, 113, 118, 189, 190, 206 228,
    233, 250, 262, 263, 268, 269
Cutler, R., 113, 119, 130
Czepiel, J. A., 48, 60

Daft, R. L., 53, 60, 108
DARPA, 188
Davis, R., 198, 209
Dearing, J. W., 17, 211, 213, 224
DEC (Digital Equipment Corporation),
    176, 178, 250
Dempsey, K., 135, 151, 159, 169
Denmark, 273
Department of Defense, 158
Deschamps, I., 49, 61
Dess, G., 84, 87
Devine, M. D., 15, 18
Diasonics, 78
Dietrich, G. B., 17, 151, 154, 157, 160,
    161, 164, 169, 171
Diffusion, 15, 34, 44, 46-49, 51-60, 80,
    104, 154, 182, 187, 221, 232, 233,
    236
Dimancescu, D., 15, 18, 95, 108
Doctors, S. I., 163, 169
Doktor, R., 50, 60
Dominguez, L., 209
Doz, Y., 64, 88
Drucker, P. F., 24, 40
Druckenberg, W. C., 156, 157, 169

EASINET (European Academic Supercom-
    puter Iniative), 181
Economics/economic development, 9, 10,
    16, 21, 22, 24-28, 32, 34, 36, 37,
    109, 111, 121, 139, 151, 153, 157,
    159, 163, 164, 167, 168, 186, 197,
    199, 207-209, 226-234, 251, 252,
    261-263
Education, 10, 12, 27, 34, 39, 114, 120,
    134, 149, 158-161, 180, 186-188,
    199, 204, 205, 214, 229, 231, 235-
    237, 248, 262-266, 268-271, 272
Egypt, 241
Elder, M., 111, 130
Electronics, 33, 55, 58, 59, 76, 86, 95, 113,
    114, 132, 133, 136, 140-143, 147-
    151, 156, 198, 201, 215, 227, 237,
    240-242, 246, 252, 253
Emery, F. E., 107, 108
Energy, 32
Ennis, D., 156, 170
Entrepreneur(ship), 13, 28-30, 37, 39, 156,
    160, 161, 164, 167, 244, 246, 247,
    255, 265
Erickson, G. A., 122, 123, 130, 158, 169
Eurich, N. P., 263, 276
Europe, 184, 187, 189, 190, 236, 267, 270
    Eastern, 38, 208
    Western, 38
European Economic Community (EEC),
    17, 226-228, 231
Evan, W, 77, 88
Eveland, J. D., 60, 61
Exchange theory, 65, 67
Expert systems, 55, 58, 97, 177
Ezzat, H. A., 272

Federal Laboratory Consortium Directory
    Resource, 155
Feigenbaum, E. P. M., 191
Fiberite Composite Materials, 139
Fisher, R., 58, 60
Fisher-Price, 200
Fishman, A. P., 164, 170
Fombrun, C., 77, 89
Ford, 72, 73, 77
Ford-Mazda, 204
Fowler, D. R., 134, 151
Frame, J. D., 259, 260, 266, 273
France, 231, 273
Frankwick, G. L., 133, 152
Friberg, E. G., 227, 238
Fry, L. W., 60, 61
Fuji, 85

Gailbraith, J. K., 22, 40
Galagan, P. A., 263, 264, 273
Garcia, C. N., 209

Gatignon, H., 48, 60
General Electric (GE), 74-78
General Motors (GM), 72, 73, 77, 78, 200, 205
Georgia Institute of Technology, 265
Gerlach, M., 85, 88
Germany, 198
Gerwin, D., 45, 59, 60
Giannisis, D., 161, 169
GIBIS (Graphical issue-based information systems), 98
Gibson, D. V., 9, 13, 15, 17, 18, 85, 86, 88, 89, 94, 96, 108, 109, 111-113, 118, 119, 153, 154, 226, 230, 231, 233, 238, 239, 247, 257, 277
Gillespie, A. E., 230, 238
Gillespie, G., 122, 123, 130, 163, 164, 169
Glaser, B. G., 98, 108
Glowasky, A. W., 164
Goddard, J. B., 230, 238
Goldstsin, M. H., 264-276
Gollub, J., 252, 256
Gomes, S., 111, 130
Gomory, R. E., 30, 33, 40
Gorbachev, M., 25, 40
Gorbis, M., 256
Gordon, D. M., 208, 209
Gore, A., Jr., 187, 191
Government, 10, 12, 28, 38, 134, 137, 142, 151, 158, 160, 163, 164, 173-175, 177, 178, 180, 182, 185, 187, 188, 190, 204, 209, 211-214, 216, 220, 228, 232, 237, 241, 245-247, 249, 250, 252-254, 262, 265
    federal, 28, 95, 111, 112, 118, 133, 138, 154, 158, 164, 196
    state, 95, 135, 136, 150, 153
Grandlin, B., 140
Great Britain, 231
Greece, 231
Grounded theory, 98
GTE, 175, 183

Hackney, S., 164, 170
Hansen, E., 256
Harcleroad, F., 11, 130
Hardware, 45, 56, 57, 59, 141, 184, 187, 189, 209, 244, 250, 255

Harrigan, K. R., 64, 69, 88
Harvard University, 16, 43
Hawley, A., 69, 88
Hazelton, J., 191
Hergert, M., 64, 88
Heterophily, 52
Hewlett Packard, 140, 181, 250
Hillenbrand, M., 111, 130
Hiltz, S., 183, 191
Hindustan Computers Limited, 248, 251
Hirsch, P., 88
Hisrich, R. D., 162, 169
History, 110-112, 121
Hitachi, 65, 74, 75, 85, 253
Hladik, K., 64, 88
Homans, G., 69, 88
Honda, 72, 73, 76, 78
Hong Kong, 241
Howell, L. J., 272
Huber, R., 230, 239
Human resources, 64

IATINET, 236
IBM, 33, 94, 140, 176, 180, 183, 248, 259, 263, 264
ICI Composites, Inc., 139
Illinois, 201, 245, 270
Incubator, 29, 156, 157, 160, 162
India, 17, 240, 241, 244, 245, 247-252, 255
Indian Institute of Science (IISc), 244
    Center for Scientific & Industrial Consultancy, 244, 245
Indian Space Research Organization (ISRO), 243, 244
Industry(ies), 10, 12, 17, 29, 32, 38, 39, 49, 63-66, 76, 109, 112-114, 116-120, 132-138, 142, 144-151, 153, 154, 158-160, 162, 164, 167, 173, 175, 177, 178, 180, 182, 185, 187, 188, 195, 196, 203-205, 207, 211, 213, 223, 227, 229, 233, 236, 240, 241, 243-249, 251-253, 259, 265, 269
Innovation (technological), 12, 16, 22, 23, 26, 29, 31, 32, 35-37, 38, 44, 46, 48, 59, 102, 153-155, 158, 161, 174, 178, 195, 198, 205, 233, 243-245, 253, 255, 263
Intel, 140

INTELSAT, 203
International, 10, 17, 22, 25, 32, 33, 38, 44,
       63, 67, 111, 153, 161, 163, 167, 174,
       175, 170, 198, 199, 209, 227, 229,
       234, 246, 248, 249, 252, 255, 258,
       259, 262, 263, 266, 268, 270, 271
Internet, 177, 178, 188
Ireland, 184, 234, 273
Italy, 17, 226, 228, 229, 231, 233, 236, 237
ITT, 74, 75
Ives, B., 49, 61

James, T. E., 15, 18
JANET (Joint Academic Network), UK,
       181
Japan, 11, 17, 24, 32, 33, 35, 38, 39, 55,
       64-66, 72, 78, 83-85, 94, 111, 118,
       119, 133, 153, 184, 187, 189, 197,
       204, 211, 212, 214, 226, 228, 250,
       253, 266
Jarillo, J. C., 86, 88
Jereski, L., 163, 169
Jerris, P., 44, 60
Joint ventures, 64, 65, 68-70, 76, 78, 85,
       135, 227
Johnson, B., 49, 61
Johnson, E. C., 148, 152
Johnston, R., 44, 60
Jones, L., 112, 130
Jones, M., 118, 130
Jordie, T., 85, 88
JUNET (Japan), 181

Kahn, R. L., 113, 130
Kamal, M. M., 276
Kanter, R., 156, 170
Kawamoto, T., 212, 215, 224
Kelley, R. W., 136, 151
Kennedy, D., 111, 130
Kentucky, 200
Killing, J. P., 64, 88
Kim, E. Y., 17, 258
Kincaid, P. L., 15, 18, 65, 67-69, 76, 80, 89,
       106, 108
Klepper, C. A., 60, 61
Kodak, 180
Kogut, B., 64, 65, 88
Korea, 197, 241, 259, 261, 267, 272, 273
Kouwenhoven, J., 112, 130

Kozmetsky, G., 15, 16, 18, 21, 85, 89, 111,
       113, 130, 131, 163, 189, 230, 233,
       239, 247, 257
Kraus, W. A., 59, 61
Krishna, S., 256
Kwiram, A. L., 113, 130

Laboratory(ies), 10-12, 17, 27, 32, 39, 60,
       154, 155, 158, 160, 163, 168, 174,
       175, 177, 180, 182, 185, 188, 205,
       213, 214, 237, 243-245
Lamont, L. M., 160, 170
Landau, R., 24, 25, 40
Langfitt, T. W., 164, 170
LaQuey, T. L., 179, 182, 191
Larsen, J. K., 133, 135, 142, 152, 156, 159,
       163, 170, 212, 213, 225, 247, 256
Laumann, E., 68, 88
Lawrence Livermore Lab, 182
Lawrence, P. R., 56, 61, 106, 108
Leadership, 21, 22, 25, 26, 28, 33, 38, 39,
       116, 148, 190, 221, 224, 253, 259
Lederberg, J., 185, 191
Legal, 85, 154
Legislation, 29, 85, 94, 96, 110, 113, 122,
       123, 138, 154, 157, 158
LeMaistre, C., 159, 168
Leonard-Barton, D., 16, 43, 49, 52, 58, 59,
       61
Lengel, R. H, 53, 60
Lesser developed countries (LDCs), 259-
       262, 265, 272
Licensing(es), 11, 33, 70, 137, 161-163,
       260
Lindsey, Q. W., 159, 160, 170
Ling, J., 123, 130
Link, A. N., 112, 122, 123, 230
Linkages, 15, 17, 23, 37, 45, 64-69, 71, 72,
       75, 76, 78, 80, 83, 84, 109, 113, 114,
       121, 164, 174, 177, 203, 207, 230,
       252
    HRIL (high resource investment link-
       age), 70
    LRIL (low resource investment link-
       age), 70
    transfer, 124-129
Lipietz, A., 209
Lockheed, 182
Lorange, P., 64, 69, 70, 87

Lorsch, J. W., 56, 61, 106, 108
Lovell, E., 64, 87
Low, G., 112, 131

MacDonald, W. B., 51, 61
Macintosh, 108
Magee, J. F., 227, 239
Maher, J., 164, 169
Malaysia, 261, 273
Management, 9, 11, 23, 26, 28, 35, 38, 44,
    47, 49, 51, 52, 55, 57-59, 66, 70, 83,
    84, 86, 93, 101, 107, 156, 162, 164,
    190, 195, 203, 205-207, 247, 255,
    259-261, 269, 270
Manhattan Project, 158
Mansfield, E., 44, 48, 61
Manufacturing, 23, 24, 30, 33, 34, 39, 45,
    48, 54, 56, 58, 60, 195-199, 201,
    203, 205-208, 213, 242, 245, 248,
    253
Maquiladora, 17, 195-206
March, J., 16, 18
Mark, H., 162, 170
Marketing, 23, 28, 65, 117, 155, 182, 186,
    228, 233, 234
Markoff, J., 112, 131
Marsden, P., 68, 85, 87, 88
Martin, E., 122, 131
Martin, P. Y., 98, 108
Massachusetts, 250
    Boston, 252
Matsushita, 74, 75
Matthai, P., 242, 256
Mazda, 72, 73
Mazey, M. E., 111, 131
McBee, F., 164
MCC (Microelectronics and Computer Tech-
    nology Corporation), 13, 17, 86, 93,
    96, 98-100, 103-105, 114, 133, 182
MCI, 175, 179
McKelvey, B., 84, 88
McLuhan, M., 13, 18
McQueen, D. H., 161, 163, 170
Medical, 32, 235, 237
Meister Engineering, 166, 167
Mendell, S., 156, 170
Merit, Inc., 179
Merrifield, B. D., 163, 170

Merton, R. K., 212, 225
Mexico, 17, 182, 183, 195-199, 201, 202,
    204, 206-209, 241, 265
Michigan, 179, 201
Michigan State University, 211
Midgley, D., 44, 61
Military, 24, 153, 158
MILNET, 178
Ministry of Trade and Industry (MITI) 66
Minor, M., 68, 87
Missouri, 201
MIT, 116, 163, 164
Mitchell, W, L., 196, 210
Mitsubishi, 65, 72, 73, 77, 85
Mitsui, 77
Mizruchi, M., 80, 88
Models, 13-16, 27, 34, 104
Monterrey Technological Institue (ITESM),
    199, 204
Mohan Ras, G. R., 249, 255
Morris, D., 64, 88
Morrison, J., 161, 170
Motivation, 12, 26, 49, 120, 164
Motorola, 85, 134
Multinationals, 10, 17, 195, 204, 205, 207,
    208, 258, 260, 261, 266, 270, 271,
    272
Mumford, E., 51, 61

Nakatani, I., 35, 40
NASA Ames Research Center, 181, 182
National Collaboratory, 185
National Research & Education Network
    (NREN), 187
National Science Foundation (NSF), 29,
    30, 40, 135, 138, 142, 148, 149,
    160, 170, 179, 180, 184
National Technological Institute (NTU),
    264, 265
NEC, 74, 75
Nepal, 261
NETT, 233-235
Network(s) 11, 15, 17, 28, 32, 37, 38, 52,
    63, 65-71, 77, 81, 84, 86, 107, 111,
    113, 114, 120, 144, 150, 156, 161,
    162, 172-190, 199-202, 205, 208,
    209, 212, 222, 223, 227, 230-236,
    252, 270, 271

analysis, 65, 67-69, 71, 73, 77, 80
Integrated Broadband Communication
    Networks, 233
ISDN, 233, 236
local area networks (LANs), 175, 177,
    188
structural equivalence, 76
wide area networks (WANs), 175, 235
New Delhi, 245, 248, 249, 253
Newly industrialized countries (NIC), 208,
    260
New York, 180, 182, 183, 200, 266
New York University, 180
NICNET (National Information Center
    Network), 252
Nii, H. P., 191
Nilekani, R., 244, 256
Nissan, 72, 73
Noble, D., 112, 131
Noling, M. S., 267, 277
NORDUNET, 189
Norling, F., 164, 168
Norris, W., 94
Nova Graphics, 166, 167
NSFNET, 177
NYSERNet, 180

Oak Ridge National Ladboratory, 159
Obermayer, H., 155, 170
OCEANIC, 189
Office of Science & Technology Policy,
    187
Office of Technology Assessment, 40
Office of Technology Business Develop-
    ment (TBD), 162
Ohmahae, K., 262, 277
Olken, H., 155, 170
Olsen, J. D., 16, 18
Olsen, M. H., 49, 61
Online Computer Library Center, Inc.
    (OCLC), 184
Organization sets, 77, 83, 84, 86, 113
Organizational environment, 16, 17, 64,
    66, 67, 69, 80, 86, 105-107, 113,
    132
Osaka University, 35
O'Shaughnessey, J., 88
Ostar, A., 111, 130

Owens, R., 110, 129

Patents, 11, 33, 137, 154, 155, 162, 163,
    198, 227
Pearson, A., 44, 60
Peck, M. J., 108
Pena, D. G., 210
Pennings, J., 69, 89
Pennsylvania Educational Communication
    Systems (PECS), 183
Pennsylvania State University, 183
Perrow, C., 45, 61
Petrella, R., 227, 239
Petrignani, R., 227, 239
Pfeffer, J., 67, 89, 113, 131
Pharmaceuticals, 35
Phillipines, 271
Piore, M. J., 197, 210
Pirkle, K., 110, 129
Pisano, G., 64, 65, 89
Pitroda, S., 245, 254
Policy(ies) 22, 24, 26, 28, 32, 112, 135,
    154, 158, 189, 190, 198, 207, 209,
    241, 246, 252, 270
Pollack, A., 30, 40
Polytechnic University, 180, 183
Porter, M., 69, 89
Portugal, 231, 273
Power, 53, 63, 65-67, 84
Powers, D. R., 159, 160, 170
Powers, M. F., 159, 160, 170
Prensky, D., 68, 88
Productivity, 22, 25, 43
Proteus, 97
Pucik, V., 64, 89
Puri, A., 256

Quaker Oats, 200
Quality, 27, 118
Quarterman, J. S., 180, 184, 191

Radian Corporation, 166, 167
Rahm, D., 154, 155, 170
Rao, K. K., 246, 256
Raytheon, 74, 75
RCA, 206
Reams, B. D., Jr., 154, 170
Reich, R., 10, 18

Rensselaer Polytechnic Institute, 164
Research, 10, 11, 14, 19, 21, 24, 27, 35, 45,
    46, 94-97, 100, 104, 111-113, 115,
    116, 119, 133-136, 138, 142, 144-
    149, 151, 153, 158-160, 162, 164,
    165, 174, 177, 178, 180-182, 184,
    186, 189, 204, 207, 211, 214-224,
    228, 233, 234, 236, 243, 245, 254
Research & Development (R&D), 10, 12,
    17, 23, 27-29, 32, 39, 48, 64-66, 69,
    78, 81, 83, 85, 93-96, 106, 112, 113,
    135, 141, 153, 154, 159, 163, 186,
    197, 198, 205, 213, 227, 240, 243-
    246, 250, 252-254, 267, 268
Research Internet Gateway (RIG), 188
Research Triangle Park, 121, 141
Resource dependence theory, 63, 65, 67,
    82, 84, 110, 120
Resources, 14, 26, 27, 30, 39, 40, 44, 59,
    64, 67, 110, 111, 120, 121, 145, 148,
    150, 155, 158, 162, 172, 173, 178,
    184, 189, 207, 232-235, 246, 252,
    262, 264, 266
Resource sharing services, 175
Rice, R., 49, 61
Rice University, 182
ROAD, 54
Roberts, E. B., 160, 170
Roberts, M., 191
Robertson, T. S., 49, 59, 61
Robinson, R. D., 277
Rochester Telephone, 180
Rockwell, Inc., 245
Roehl, T., 85, 89
Rogers, D. M. A., 33, 40
Rogers, E. M., 13, 15, 17, 18, 22, 44, 48,
    49, 52, 59-61, 65, 67-69, 76, 80, 86,
    88, 89, 94, 108, 112, 130, 133, 135,
    142, 152, 156, 159, 163, 170, 212,
    240, 241, 245, 247, 252, 256
Rosenbloom, R., 111, 129
Rostow, W. W., 22, 30, 38, 40, 153, 171
Rousseau, D. M., 45, 62
Rubenstein, A. H., 44, 62
Russo, M., 64, 65, 89
Ryans, J. K., 159, 163, 171

Sabatelli, R., 231, 238

Sabel, D. F., 197, 210
Safa, P., 209
Salnacik, G., 67, 89, 113, 131
Sanger, D., 89
Sassen, S., 208, 210
Sawhney, H., 17, 240, 252, 256
SCHEDULER, 54
Schmitt, R. W., 30, 33, 40
Schramm, W., 13, 18
Schlumberger, 182
Schmandt, J., 186, 191
Schon, D. A., 62
Schultz, R. L., 50, 60
Scotland, 184
Security, 2, 27
SEMATECH, 86, 114, 133, 182
Semiconductors, 26, 86, 117, 158
SemiMAC, 141
Sesquinet, 182
Sexton, D. L., 163, 171
Shan, W., 64, 89
Shanklin, W. L., 159, 163, 171
Shapero, A., 157, 171
Silicon Valley, 121, 140, 230, 248, 252
SIM, 54
Sims, R. R., 111, 131
Singapore, 241, 271, 273
Singh, A., 64, 65, 88
Singhal, A., 17, 240, 241, 245, 247, 256
Sklair, L., 196, 198, 210
Slevin, D. P., 50, 60
Smilor, R. W., 13, 15, 17, 85, 89, 93, 94,
    96, 108, 111, 113, 118, 119, 129
    130, 131, 162-164, 169, 171, 230,
    231, 233, 238, 239, 247, 257
Smith, K. K., 98, 107
SNAIL, 58
Snowdon, M., 227, 239
Snyder, D. R., 133, 134, 152
Software, 35, 45, 54-56, 58, 59, 97, 140,
    141, 156, 177, 181, 184-187, 189,
    202, 203, 209, 241, 244, 246, 248-
    253, 270
Sokol, L., 157, 171
Sony, 74, 75
Souder, W., 44, 62
Spain, 231, 273
Spin-outs, 17, 153-157, 160, 164, 167

SRI International, 181
Sri Lanka, 261
Stanford Research Park, 141
Stanford University, 116, 117, 161, 163, 164, 181, 265
Stanley, D., 131
STAR Special Telecomnunications Action for Regional Development, 231, 233, 236
Starling, B., 140
Starlink, 184, 189
Starr, M. K., 34, 40
State University of New York at Buffalo, 180, 181
Steinnes, D. N., 159, 171
Stewart, G. H., 17, 115, 116, 118
Siffler, A., 256, 277
Strategy, 11, 24, 25, 39, 54, 57, 64, 66, 78, 81, 84, 100, 103, 143
Strauss, A. L., 98, 108
Strizich, M., 277
Subramanya M., 245, 257
Sudershan, P., 247
Superconductivity, 24, 38, 86, 98, 112, 173
Superconductivity Super Collider Laboratory, 182
Susbauer, J. C., 159, 160, 171
Szgenda S., 161, 162, 171

Taiwan, 201, 241, 261
Tatsuno, S., 182, 191, 239
Technology, 23-25, 27, 37, 45, 48, 50, 53, 55, 60, 97, 98, 106, 112, 114-116, 158, 162, 182, 187, 197, 198, 203, 227, 229, 235, 251-253, 255, 258-264, 266, 269, 270-271
    characteristics, 38, 102
    commercialization, 16, 21-23, 26-31, 37, 39, 44, 112, 113, 120, 153-155, 159, 161, 162, 168, 173, 246, 253, 254
    customization, 47, 55, 56
    development, 33, 51
    stages, 14, 22, 51
Technology Development & Information Company of India (TDICI), 247, 248
Technology Futures, Inc., 93

Technology transfer, 9-12, 15, 16, 21, 33, 43, 44, 46, 48, 49, 59, 64, 70, 76, 80, 84, 93-95, 99, 103, 104, 106, 109, 113, 114, 116, 118, 132, 133, 138, 120, 141, 143, 146, 149, 150, 155, 160-163, 168, 172-174, 180, 181, 186, 189, 190, 195, 196, 198, 199, 204-209, 212-215, 222-224, 227, 231, 240, 241, 243-249, 251-255, 258-261, 265-267, 270-272
    definitions, 11, 13, 45, 106, 141, 142
    factors, 30-31
    mechanisms, 100-103
    reverse, 248, 255
    success, 12, 14, 26, 119
Technopolis(eis), 17, 32, 85, 111, 113, 121, 182, 226, 228-230, 233, 246, 247, 249, 255
Tecnopolis Novus Ortus (TCNO), 17, 226, 228-231, 233, 234, 236
Teece, D., 64, 65, 85, 88, 89
Telecommunications, 11, 17, 38, 65, 76, 135, 136, 172-175, 183, 186, 189, 190, 195, 197, 199, 203, 205-209, 226, 227, 230-233, 241, 245, 246, 252-254, 269, 272
TELMEX, 200-202
Terpstra, V., 111, 131
Terreberry, S., 107, 108
Texas, 199, 201, 203, 251
    Austin, 17, 86, 97, 113, 116, 121, 167, 198, 199
    San Antonio, 141
Texas A&M University, 162, 182
Texas Instruments (TI), 85, 182, 248-251, 255
Thackery, R., 110, 131
THEnet (Texas Higher Education Network), 182
Thorelli, H., 86, 89
Tichy, N., 77, 89
Tillman, S. A., IV, 113, 123, 129, 158, 169
Tornatzky, L. G., 148, 152
Toshiba, 65, 74-78, 83
Toyota, 72, 73
Tracor Inc., 164, 166, 167
Training, 27, 70, 204, 206, 231, 234-237, 251, 258, 260, 262, 263, 266-272

Transaction cost approach, 63, 65, 66, 81, 84
Transamerica Research Center, 139
Trist, E. L., 107, 108
Truitt, J. F., 85, 89
Tsukuba Science City, 17, 211-219, 221-224
Turoff, M., 183, 191
Turner, B. A., 98, 108
Tushman, M., 77, 89
Tyson, L., 34

Ulrich, R. A., 45, 62
Uncapher, K., 185, 191
United Kingdom, 184, 251, 270
United States, 11, 17, 22-24, 27-29, 33, 35, 37, 39, 55, 64, 66, 72, 75, 77, 83-86, 94-96, 109, 110, 112, 113, 118-121, 133, 137, 153, 154, 156-160, 168, 174, 182, 187, 189, 190, 196-199, 201, 203, 205, 206, 208, 209, 216, 248-250, 252-254, 259, 260, 265, 266
University of Bari (Italy), 228
University of California, 181
    Berkeley, 34, 137
University of Chicago, 155, 158
University of Houston, 116
University of Southern Califonia, 9, 212, 241
University of Texas:
    Austin, 7, 9, 11, 16, 17, 21, 25, 93, 116, 117, 161, 162, 167, 241
    El Paso, 205
    San Antonio, 154, 199
University of Wisconsin, 163
University(ies), 10, 12, 17, 27, 28, 35, 39, 95, 96, 109-122, 132, 133, 135, 137-139, 142, 144-151, 157-164, 167, 168, 174, 175, 177-185, 187, 188, 204, 213, 214, 216, 219, 220, 228, 229, 244, 251
UNIX, 248
Ury, W., 58, 61
Users, 15, 45, 47, 49, 50-55, 57, 59, 141-143
    involvement, 49-52
US Sprint, 175

USSR, 25, 26, 32, 38, 84
Value(s), 16, 23, 26, 118, 121, 161, 164, 222, 268
Vargas, L., 196, 210
Varma, K., 242, 257
Ventures, 17, 28, 66, 70, 112, 137, 153, 154, 156, 161, 162, 189, 246, 248, 250
VLSI Technology Inc., 140, 141
von Hippel, E., 50, 62
Vyasulu, V., 242, 257

Wacholder, M., 159, 168
Waldhorn, S., 256
Wales, 184
Walker, G., 86, 89
Wallace, G., 113, 123, 130
Wallmark, J. T., 161, 163, 170
Walton, R. E., 49, 62
Warren, R., 85, 89
Washington, DC, 252
Waugaman, P. G., 122, 131
Weber, M., 212, 225
Weick, K., 13, 18, 106, 108
Weiland, G. F., 45, 62
Weinberg, M., 111, 131
Weintraub, S., 22, 40
Weiss, G., 155, 171
Wetzel, W., 161, 170
Whetten, D., 67, 87
Whitney, D. E., 50, 62
Wigand, R. F., 17, 132, 133, 135, 142, 152, 161, 171
Wilem, F., 159, 171
Williams, F., 9, 17, 103, 108, 172, 186, 195, 196, 277
Williamson, O., 89
Willis, R., 161, 169
Wilson, M., 161, 171, 186
Wilson, P., 198, 208, 210
Wohlert, K. L., 13, 15, 17, 18, 277, 278
Woodward, J., 45, 62
Wulf, W. A., 179, 185, 191

Xerox, 181

Zenith Electronics Coporation, 201, 202
ZYCOR, 166, 167

NOTES

NOTES